Lecture Notes
in Control and Information Sciences   227

Editor: M. Thoma

Springer
London
Berlin
Heidelberg
New York
Barcelona
Budapest
Hong Kong
Milan
Paris
Santa Clara
Singapore
Tokyo

Sophie Tarbouriech and Germain Garcia (Eds)

Control of Uncertain Systems with Bounded Inputs

Springer

Series Advisory Board

A. Bensoussan · M.J. Grimble · P. Kokotovic · H. Kwakernaak
J.L. Massey · Y.Z. Tsypkin

Editors

Sophie Tarbouriech
Germain Garcia
LAAS-CNRS, 7 Avenue du Colonel Roche, 31077 Toulouse Cedex 4, France

ISBN 3-540-76183-7 Springer-Verlag Berlin Heidelberg New York

British Library Cataloguing in Publication Data
Control of uncertain systems with bounded inputs. -
  (Lecture notes in control and information sciences ; 227)
  1.Feedback control systems 2.Uncertainty (Information
  theory)
  I.Tarbouriech, Sophie II.Garcia, Germain
ISBN 3540761837

Library of Congress Cataloging-in-Publication Data
Control of uncertain systems with bounded inputs / Sophie Tarbouriech
  and Germain Garcia, eds.
     p.   cm. - - (Lecture notes in control and information sciences
  ; 227)
  ISBN 3-540-76183-7 (pbk. : alk. paper)
  1. Real-time control. 2. Systems analysis. I. Tarbouriech,
  Sophie. II. Garcia, Germain. III. Series.
  TJ217.7.C66 1997   97-14914
  629.8'312- -dc21   CIP

Typesetting: Camera ready by editors
Printed and bound at the Athenæum Press Ltd, Gateshead
69/3830-543210 Printed on acid-free paper

Preface

In a practical control problem, many constraints have to be handled in order to design controllers which operate in a real environment. The first step in a control problem is to find an appropriate model for the system. It is well-known that this step, if successfully applied (that is, if the model gives an accurate representation of physical phenomena) leads usually to a satisfactory control law design, satisfactory meaning that the observed behavior of the real controlled system is conform with the desired results.

A model can be derived in several ways. The most direct approach consists in applying general physical laws, decomposing the modelling problem into subproblems, solving each of them and by some more or less simple manipulations, deriving a complete model. Obviously, this method is based on some strong a priori knowledges and then some approximations are usually considered.

For some systems, the previous approach is difficult to implement because the application of physical laws is practically impossible or simply because only a partial knowledge on the system is available. In this case, the system is considered as a black box and a model is elaborated from experimental data (identification). Some crucial choices have to be done in order to derive a satisfactory model, these choices concerning essentially the input, the model order and model structure.

It is also possible to combine the two previous methods. An a priori knowledge on the system is then combined with identification tools. The model structure results from the a priori knowledge while the model parameters are obtained by an identification method.

In conclusion, to obtain a model using one of the above approaches, it is often necessary to approximate or neglect some phenomena, or to choose some key parameters. A direct consequence is that the derived model is affected by some uncertainties. To find a control operating in a real environment, uncertainties have to be appropriately described and their effects considered in the control law design (Robust Control). Some potential results on robust control have been widely developed these last decades and although, some intensive works continue to be developed, this domain has attained a certain maturity degree.

Concerning the control, practically the control is bounded and saturations can occur, these problems being the consequence of actuators limitations. It is also important to include them in the control law design (Constrained Control). A large amount of works has been done in this way and several approaches were developed in the literature.

It seems to be fundamental to combine the results obtained in these two fields in order to derive some methodologies with practical interest and therefore to design some controllers capable of achieving acceptable performances under uncertainty or disturbance and design constraints. For two or three years, a significative effort is done in this sense. This book entitled "*Control of Uncertain Systems with Bounded Inputs*" is aimed to give a good sample, not exhaustive of course, of what it was done up to now in the field of robust and constrained control design. The idea is to propose a collection of papers in which some fundamental ideas and concepts are proposed, each paper constituting a chapter of the book. The book is organized as follows.

Chapter 1. Feedback Control of Constrained Discrete-Time Systems
by E. De Santis

In this chapter, the author considers a linear discrete-time system with exogenous disturbances and with bounded inputs and states. The problem of controlling such a system is addressed. Conditions are proposed for the existence of state feedback controllers that achieve a level of performance with respect to some given criteria. An additional requirement is that the undisturbed system is asymptotically stable. If the problem has solution, the subclass of controllers that achieve the highest level of robustness, with respect to given parametric uncertainties in the system matrices, is determined.

Chapter 2. $\mathcal{L}^2$-Disturbance Attenuation for Linear Systems with Bounded Controls : an ARE-Based Approach
by R. Suárez, J. Alvarez-Ramírez, M. Sznaier, C. Ibarra-Valdez

Linear continuous-time systems with additive disturbance and bounded controls are considered. A technique for obtaining a bounded continuous feedback control function is proposed in order both to make globally stable the closed-loop system and to satisfy an $\mathcal{L}_2$ to $\mathcal{L}_2$ disturbance attenuation in a neighborhood of the origin. The solution is given in terms of solutions to an algebraic parametrized Riccati equation. The proposed control is then a linear-like feedback law with state-dependent gains.

Chapter 3. Stability Analysis of Uncertain Systems with Saturation Constraints
by A.N. Michel and L. Hou

This chapter addressed new sufficient conditions for the global asymptotic stability of uncertain systems described by ordinary differential equations under saturation constraints. Systems operating on the unit hypercube in $\Re^n$ (where all states are subject to saturation constraints) and systems with partial state saturation constraints (where only some of the states are subject to constraints) are studied. These types of systems are widely used in several areas of applications, including control systems, signal processing, and artificial neural networks. The usefulness of the proposed results is shown by means of a specific example.

Chapter 4. Multi-Objective Bounded Control of Uncertain Nonlinear Systems : an Inverted Pendulum Example
by S. Dussy and L. El Ghaoui

Considering a nonlinear parameter-dependent system, an output feedback controller is sought in order for the closed-loop system to satisfy some specifications as stability, disturbance rejection, command input and output peak bounds. The controller state space matrices are allowed to depend on a set of measured parameters and/or states appearing in the nonlinearities. The specifications have to be robustly satisfied with respect to the remaining (unmeasured) parameters and/states appearing in the nonlinearities. Sufficient conditions are derived to ensure the existence of such a mixed gain-scheduled/robust controller. These conditions are LMIs, associated with a set of nonconvex conditions. An efficient heuristic to solve them is proposed. The design method is illustrated with an inverted pendulum example.

Chapter 5. Stabilization of Linear Discrete-Time Systems with Saturating Controls and Norm-Bounded Time-Varying Uncertainty
by S. Tarbouriech and G. Garcia

Discrete-time systems with norm-bounded time-varying uncertainty and bounded control are considered. From the solution of a discrete Riccati equation a control gain and a set of safe initial conditions are derived. The asymptotic stability of the closed-loop system is then locally guaranteed for all admissible uncertainties. The connections between these results and the disturbance rejection problem are investigated. The class of perturbations which can be rejected in presence of saturating controls is characterized. The results are illustrated with the discretized model of the inverted pendulum. Furthermore, the approach by LMIS is discussed.

Chapter 6. Nonlinear Controllers for Constrained Stabilization of Uncertain Dynamic Systems
by F. Blanchini and S. Miani

The problem of determining and implementing a state feedback stabilizing control law for linear continuous-time dynamic systems affected by time-varying memoryless uncertainties in the presence of state and control constraints is addressed. The properties of the polyhedral Lyapunov functions, i.e. Lyapunov functions whose level surfaces aretheir capability of providing an arbitrarily good approximation of the maximal set of attraction, which is the largest set of initial states which can be brought to the origin with a guaranteed convergence speed. First the basic theoretical background needed for the scope is recalled. Some recent results concerning the construction of the mentioned Lyapunov functions and the controller implementation are reported. Finally, the results of the practical implementation on a two-tank laboratory system of a linear variable-structure and a quantized control law proposed in literature is presented.

Chapter 7. H_∞ Output Feedback Control with State Constraints
by A. Trofino, E.B. Castelan, A. Fischman

In this chapter, a biconvex programming approach is presented for the design of output feedback controllers for discrete-time systems subject to state constraints and additive disturbances. The method proposed is based on necessary and sufficient conditions for the existence of stabilizing static output feedback controllers. Mixed frequency and time domain specifications for the closed-loop system like H_∞ performance requirements and state constraints in the presence of disturbances are investigated.

Chapter 8. Dynamic Output Feedback Compensation for Systems with Input Saturation
by F. Tyan and D.S. Bernstein

This chapter deals with optimization techniques to synthesize feedback controllers that provide local or global stabilization along with suboptimal performance for systems with input saturation. The approach is based upon LQG-type fixed-structure techniques that yield both full and reduced-order, linear and nonlinear controller. The positive real lemma provides the basis for constructing nonlinear output feedback dynamic compensators. A major aspect of the presented approach is the guaranteed subset of the domain of attraction of the closed-loop system. The results are then illustrated with several numerical examples.

Chapter 9. Quantifier Elimination Approach to Frequency Domain Design
by P. Dorato, W. Yang, C. Abdallah

Quantifier-elimination methods are proposed for the design of fixed structure compensators which guarantee robust frequency domain bounds. It is shown for example, that robust stability and robust frequency domain control-effort constraints, can be reduced to system of multivariable polynomial inequalities, with logic quantifiers on the frequency variable and plant-parameter variables. Quantifier-elimination software can then be used to eliminate quantifier variables and to obtain quantifier-free formulas which define sets of admissible compensators parameters.

Chapter 10. Stabilizing Feedback Design for Linear Systems with Rate Limited Actuators
by Z. Lin, M. Pachter, S. Banda, Y. Shamash

This chapter considers two design techniques recently developed for linear systems with position limited actuators : piecewise-linear LQ control and low-and-high gain feedback. These techniques are combined and applied to the design of a stabilizing feedback controller for linear systems with rate-limited actuators. An open-loop exponentially unstable F-16 class fighter aircraft is used to demonstrate the efficiency of the proposed control design method. The proposed combined design takes advantages of these techniques while avoiding their disadvantages.

We hope that this book would highly contribute to significative developments in the field of robust and constrained control, and would be a reference for future investigations in this field.

Toulouse, April 8, 1997.

Sophie Tarbouriech   Germain Garcia
L.A.A.S.-C.N.R.S.   L.A.A.S.-C.N.R.S.
I.N.S.A.T

Table of Contents

List of Contributors

Siva Banda
Flight Dynamics Dr. (WL/FGIC)
Wright Laboratory
Stony Brook, NY 11794-3600, USA.

Franco Blanchini
Dipartimeto di Matematica e Informatica,
Universita Degli studi di Udine,
Via Zannon 6, 33100 Udine, Italy.
E-mail : blanchini@uniud.it

Denis S. Bernstein and Feng Tyan
Department of Aerospace Engineering,
The University of Michigan,
Ann Arbor, MI 48198, USA.
E-mail : dsbaero@engin.umich.edu

Elena De Santis
University of L'Aquila,
Department of Electrical Engineering,
67040 Poggio di Roio (L'Aquila), Italy.
E-mail : desantis@dsiaq1.ing.univaq.it

Peter Dorato, Wei Yang and Chaouki Abdallah
Department of Electrical and Computer Engineering,
University of New Mexico,
Albuquerque, NM 87131-1356, USA.
E-mail : peter@jemez.eece.unm.edu

Stéphane Dussy and Laurent El Ghaoui
Laboratoire de Mathématiques Appliquées,
Ecole Nationale Supérieure Techniques Avançées,
32 Boulevard Victor, 75739 Paris, France.
E-mail : dussy@ensta.ensta.fr

Arão Fischman
Laboratoire d'Automatique de Grenoble (URA CNRS 228),
ENSIEG, BP 46, 38402 St.-Matind'Hères, France.

Zongli Lin
Dept. of Applied Math. & Stat.
SUNY at Stony Brook
Stony Brook, NY 11794-3600, USA.
E-mail : lin@ams.sunysb.edu

Stefano Miani
Dipartimento di Elettronica e Informatica,
Università degli Studi di Padova,
Via Gradenigo 6/a, 35131 Padova, Italy.
E-mail : miani@dimi.uniud.it

Anthony N. Michel and Ling Hou
Department of Electrical Engineering,
University of Notre Dame,

Notre Dame, IN 46556, USA.
E-mail : anthony.n.michel.l@nd.edu

Meir Pachter
Dept. of Elect. & Comp. Sci.
Air Force Inst. of Tech.
Wright-Patterson, AFB, OH 45433,
USA.

Yacov Shamash
College of Engr. & Applied Sci.
SUNY at Stony Brook
Stony Brook, NY 11794-2200, USA.

Rodolfo Suárez, Josè Alvarez-Ramirez, C. Ibarra-Valdez
División de Ciencias Básica e Ingeniería,
Universidad Autónoma Metropolitana
- Iztapalapa,
Apdo Postal 55-534, 09000 México
D.F., México.
E-mail : rsua@xanum.uam.mx

Mario Sznaier
Department of Electrical Engineering,
The Pennsylvania State University,
University Park, PA 16802, USA.
E-mail : msznaier@frodo.ee.psu.edu

Sophie Tarbouriech and Germain Garcia
LAAS-CNRS
7 avenue du colonel Roche, 31077
Toulouse cedex, France.
E-mail: tarbour@laas.fr

Alexandre Trofino and Eugenio Castelan
Laboratório de Controle e Microinformática (LCMI/EEL/UFSC)
Universidade Federal de Santa Catarina
PO 476, 88040-900,

Florianópolis (S.C.), Brazil.
E-mail : trofino@lcmi.ufsc.br

Chapter 1. Feedback Control of Constrained Discrete-Time Systems*

Elena De Santis

Department of Electrical Engineering, University of L'Aquila, Monteluco di Roio, 67040 L'Aquila. FAX+39-862- 434403. E-mail: desantis@dsiaq1.ing.univaq.it

1. Introduction and problem statement

Let us consider the system

$$x(t+1) = Ax(t) + Bu(t) + D\delta(t) \tag{1.1}$$

where the vectors $x(t) \in R^n$, $u(t) \in R^m$ and $\delta(t) \in R^d$ represent respectively the state, the input and the disturbance, and let us consider the sets $X_0 \subseteq R^n$, $\Sigma \subseteq R^n$, $U \subseteq R^m$ and $\Delta \subseteq R^d$, where X_0 is bounded, $0 \in \Delta$, $0 \in X_0$, $0 \in \Sigma$, and $X_0 \subseteq \Sigma$.

Let us define the following problem:

PROBLEM (P1):
Given a set $\Pi' \subseteq R_+^2$, where R_+^2 denotes the nonnegative orthant of the space R^2, find a control law and two parameters γ and $\rho \in \Pi'$ such that:
a) $u(t) \in \rho U$ $\forall t$
b) $\forall x_0 \in X_0$, $x(t) \in \Sigma$, $\forall \delta(t) \in \frac{\Delta}{\gamma}$, $\forall t > 0$
c) if $\delta(t) = 0$, $t \geq \bar{t}$, the evolution starting from $x(\bar{t})$ tends to the origin with some rate of convergence, $\forall x(\bar{t}) \in X_0$.

The theory of positive invariant sets plays a central role in our approach. Therefore we recall now the main definitions:

Definition 1.1. *Given a set V and the constraint $u \in V$, a set X with $0 \in X$ is λ-contractive controllable (shortly λ - c.c.) if $\forall x \in X, \exists u \in V : Ax + Bu \in \lambda X$. If $\lambda=1$, the set X is positively invariant controllable (shortly p.i.c.).*

Definition 1.2. *Given a set V and the constraint $u \in V$, a set X is Δ-positive invariant controllable (shortly Δ- p.i.c) if $\forall x \in X, \exists u \in U : Ax + Bu + D\Delta \subseteq X$.*

Problem (P1) above can be formulated as a problem of existence of some set. To this aim we can state the following proposition:

---

* Work supported by MURST-40 and MURST-60

Proposition 1.1. *Problem (P1) has a solution if the following problem (P2) has a solution:*

Given the set $\Pi = \Pi' \times \{\lambda : 0 \leq \lambda < 1\}$, find a triple $(\lambda, \gamma, \rho) \in \Pi$, and a bounded set $\Sigma(\lambda, \gamma, \rho)$ such that:

i) $X_0 \subseteq \Sigma(\lambda, \gamma, \rho) \subseteq \Sigma$

ii)$\forall x \in \Sigma(\lambda, \gamma, \rho) \; \exists u \in \rho U : Ax + Bu \in \lambda\Sigma(\lambda, \gamma, \rho)$ and $Ax + Bu + \frac{D\Delta}{\gamma} \subseteq \Sigma(\lambda, \gamma, \rho)$

Given a triple $(\lambda, \gamma, \rho) \in \Pi$, denote by $X_{\lambda\gamma\rho}$ the family of all the sets $\Sigma(\lambda, \gamma, \rho)$ with properties i and ii defined in Proposition 1.1.

Lemma 1.1. *The family $X_{\lambda\gamma\rho}$, if nonempty, has a maximal element*

$$\Sigma^*(\lambda, \gamma, \rho).$$

Proof: It suffices to notice that if $X_1 \in X_{\lambda\gamma\rho}$ and $X_2 \in X_{\lambda\gamma\rho}$, the set $X_1 \cup X_2$ belongs to the same family $X_{\lambda\gamma\rho}$. $\square$

A number of subproblems can be formulated, relaxing or particularizing our requirements. We cite now only some main examples:

i) if $\Sigma = X_0$, $\Delta = \{0\}$ and $\rho = 1$ the problem reduces to the problem studied in [12].

ii) if $U = R^m$, $X_0 = \{0\}$, $\Delta = \{\delta : \|\delta\|_\infty \leq 1\}$, $\Sigma = \{x : \|x\|_\infty \leq \gamma^*\}$, where γ^* is the minimum γ such that $\|x(.)\|_\infty \leq \gamma\|\delta(.)\|_\infty$, $x(0) = 0$, Problem (P1) reduces to a classical l_1 optimal control problem (see e.g. [3]). We recall that, as it was shown in [9], even in the case of full state feedback control, l_1 optimal and near optimal linear controllers may be dynamic and of arbitrarily high order. In [14] a constructive algorithm is presented for the computation of near optimal nonlinear controllers which are static. Our approach goes in this direction, but while the above cited algorithm recursively increase the performance, until a value close to the optimal performance is obtained, our technique may be considered "one-shot", in a sense that will be clear later.

Obviously, if U is some set in R^m we have an input constrained l_1 problem.

iii) if $X_0 = \{0\}$, $U = R^m$, $\gamma = \bar{\gamma}$, $\lambda = \bar{\lambda}$ problem (P2) reduces to the problem studied in [10].

iv) if $\Sigma = R^n$, $\Delta = \{0\}$, $\rho = 1$, X_0 is a given but arbitrarily large set, the problem becomes a problem of constrained exponential stabilization as studied e.g. in [13].

v) if $\Delta = \{0\}$, $\lambda = \bar{\lambda}$, $\rho = 1$ and X_0 is to be determined, the problem is that of finding the largest set in Σ, such that for each initial point in the set, the evolution exponentially converges to the origin, with assigned rate of convergence $\bar{\lambda}$, being the input constrained in U.

vi) If $\Delta = \{0\}$, and if, for a given $\lambda = \bar{\lambda}$, it is required to minimize the parameter ρ in such a way that conditions i) and ii) in proposition 1.1 are fulfilled, the problem is that of locally stabilize the system, with prescribed rate of convergence, with minimum effort.

The problem studied in [1] is a variant of problem (P1). In fact in that paper the problem was the following: for a given set $X_0 = \Sigma$ and a given λ, find a set Σ_λ and a control law, such that $\Sigma_\lambda \in \Sigma, \forall x \in \Sigma_\lambda \exists u \in U :$ $Ax + Bu + D\Delta \subseteq \lambda\Sigma_\lambda$. In the same paper this problem was solved also in the case of matrices A and B dependent on unknown parameters, bounded in a known set.

If we analyze Problem (P2), we can say that, in general, if the problem has solution, this solution is not unique. In fact we have a set $\Phi \in \Pi$ of feasible values (λ, γ, ρ), i.e. of values such that a set $\Sigma(\lambda, \gamma, \rho)$ with properties defined in the formulation of Problem (P2) exists. Therefore, our objective is not only to find a solution of P2, but also to optimize our choice with respect to some given criteria. Let us call P3 the class of these optimization problems.

In this respect, the main characteristic of our approach is that, choosing a set Π as large as possible, (which means setting initial requirements as loose as possible), we use a technique which allows to obtain a control law and values of the parameters in the set Φ, starting from feasible values $(\lambda_0, \gamma_0, \rho_0)$, optimizing with respect to some criteria. This in substance means that the performance of the controlled system is not a priori fixed, but the choice of a satisfactory level of the parameters λ and/or γ and/or ρ is case driven, taking also into account numerical limitation.

As for the criteria to choose a solution in the set Φ, the idea is that of associating to each parameter a priority: e.g., for a fixed $\rho = \bar{\rho}$, the problem is that of obtaining a satisfactory level of contractivity factor λ and for that choice one wants to minimize the parameter γ. In this example we have associated to the parameter ρ the highest priority (it is the case in which hard constraints are to be considered for the input); we have associated to the parameter γ the lowest priority (this means that the level of the disturbances in the system is not critical). In this case $\Pi = \{\lambda, \gamma, \rho : 0 \leq \lambda < 1; 0 \leq \gamma; 0 \leq \rho \leq \bar{\rho}\}$. With this setting obviously it is well possible that the problem has not solution, but, as we'll see later in section 4, this in general does not imply a complete reinitialization of the algorithm.

In contrast to our approach, notice that in [10] and in [1], respectively the values $(\bar{\lambda}, \bar{\gamma})$ and $\bar{\lambda}$ are a priori fixed and, if the problem cannot be solved with those choices, new values of $\bar{\lambda}$ and/or $\bar{\gamma}$ are assumed, reinitializing the algorithm and iterating the procedure until feasible values are found. A similar remark applies also to [14].

In this paper we make the assumptions that X_0, Σ and U have nonempty interior, the origin belongs to the interior of each of them and X_0 and Σ are bounded. Notice that the assumption that U has interios implies no loss of generality. In fact, it this was not the case, it should be always possible to compute a matrix $\tilde{B}$ and a set $\tilde{U}$ such that $\tilde{B}\tilde{U} = BU$, and $\tilde{U}$ has nonempty interior. The assumptions on X_0 and Σ are made to assure the stability of the closed loop system.

Under these assumptions, a necessary condition under which Problem P2 has solution is that system (1.1) is stabilizable. This condition is equivalent to say that system (1.1), with $u(t) \in \rho U, \forall t \geq 0$, for all positive values of parameter ρ, admits a bounded λ-c.c. convex set, with the origin in its interior, for some $\bar{\lambda}$, $0 \leq \bar{\lambda} < 1$. This implies that system (1.1), with $u(t) \in \rho U, \forall t \geq 0, \forall \rho > 0$ admits a bounded λ-c.c. polytope, with the origin in its interior, for all values of parameter λ, such that $\bar{\lambda} < \lambda \leq 1$ (see [1], theorem 3.2).

If Σ and U are convex polytopes, we are able to explicitly compute a triple $(\lambda, \gamma, \rho) \in \Phi$ and the set $\Sigma^*(\lambda, \gamma, \rho)$.

We'll see that if Σ and U are polyhedra, in general the controls that solve a problem in the class (P3) are static nonlinear state feedback controls. In fact, given a problem, we'll show that $u(t)$ is a solution, if and only if: $Vu(t) \leq f(x(t), \lambda, \gamma, \rho)$, $t \geq 0$, where V is a constant matrix, depending on the the matrices A and B of the dynamical system and on the coefficient matrices defining the polyhedra Σ and U. The vector $f(x(t), \lambda, \gamma, \rho)$ is the bound vector, which, for fixed values of the parameters λ, γ and ρ depends on the state vector at time t.

We make use of a duality result, which will be recalled now for reader's convenience. Let G be a $s \times n$ matrix and let v be a vector in R^n. Denote by $\mathcal{R}(G)$ the range of the matrix G and by P the nonnegative orthant of R^s, i.e., for simplicity, $P = R_+^s$. With a slight abuse of notations, in the sequel we'll denote the nonnegative orthant of any space with the same symbol P. The convex cone $\mathcal{R}(G)^\perp \cap P$ is polyhedral and pointed. A set of vectors formed taking a nonzero vector from each of its extreme rays is called set of generators of the cone.

At this point we can state the following

Theorem 1.1. *[4] The convex polyhedron $\{x : Gx \leq v\}$ is nonvoid in and only if*

$$Qv \geq 0$$

where

$$\{\text{rows of } Q\} = \{\text{generators of the cone} : \mathcal{R}(G)^\perp \cap P\}.$$

Given some linear programming problem, it can be solved as a parameterized feasibility problem as follows:

$$\max_{x:Gx\leq v} (f,x) = \max h : Q \begin{pmatrix} v \\ -h \end{pmatrix} \geq 0 \qquad (1.2)$$

where

$$\{\text{rows of } Q\} = \left\{ \text{generators of the cone} : \mathcal{R} \begin{pmatrix} G \\ -f \end{pmatrix}^{\perp} \cap P \right\}$$

In [5] this conical approach to linear programming is extensively studied and moreover an implementation in MODULA-2 of an efficient algorithm for the computation of the generators of the cone $\mathcal{R}(G)^{\perp} \cap P$ is given. We'll see in the sequel that the computation of these generators plays a central role in our approach.

The paper in organized as follows: in the section 3 we study some properties of a set in the family $X_{\lambda\gamma\rho}$ and we describe a technique to compute the set $\Sigma^*(\lambda, \gamma, \rho)$, given the parameters (λ, γ, ρ).

In section 4 we give the tools to solve problem (P3). A numerical example is also developed.

Finally in section 5 we address the problem of evaluating the degree of robustness of a given solution, with respect of a given description of the uncertainties in the system matrix A.

2. Preliminary results

In the following theorem we study how some properties of a set $X \in X_{\lambda\gamma\rho}$ depend on properties of the set BU and of the set Δ. The set Φ, as we have defined in the previous section, is the set of fesible parameters in Π. The symbol $\mathcal{C}(X)$ denotes the convex hull of some set X

Theorem 2.1. *Given a triple* $(\lambda, \gamma, \rho) \in \Phi$ *and a set* $X \in X_{\lambda\gamma\rho}$ *we have:*
  i) if BU *and* Σ *are convex,* $\mathcal{C}(X) \in X_{\lambda\gamma\rho}$
  ii) if BU *and* Σ *are convex and 0-symmetric,* $\mathcal{C}(X \cup (-X)) \in X_{\lambda\gamma\rho}$.

Proof:
  i) If $X \in X_{\lambda\gamma\rho}$ we can write:

$$\forall x_1 \in X \; \exists u_1 \in \rho U : Ax_1 + Bu_1 \in \lambda X \text{ and } Ax_1 + Bu_1 + D\frac{\delta_1}{\gamma} \in X, \; \forall \delta_1 \in \Delta$$

$$\forall x_2 \in X \; \exists u_2 \in \rho U : Ax_2 + Bu_2 \in \lambda X \text{ and } Ax_2 + Bu_2 + D\frac{\delta_2}{\gamma} \in X, \; \forall \delta_2 \in \Delta$$

therefore $\exists u_1, u_2 \in BU$ such that :

$$A(\alpha x_1 + \beta x_2) + w \in \lambda \mathcal{C}(\mathcal{X})$$
$$\text{and}$$
$$A(\alpha x_1 + \beta x_2) + (\alpha B u_1 + \beta B u_2) + D\frac{\mathcal{C}(\Delta)}{\gamma} \in \mathcal{C}(X)$$

$$x_1, x_2 \in X, \ \forall \alpha, \beta : \alpha + \beta = 1, \ \alpha \geq 0, \ \beta \geq 0$$

and hence, because $X_0 \subseteq \mathcal{C}(X)$, $\mathcal{C}(X) \subseteq \Sigma$ and $\Delta \subseteq \mathcal{C}(\Delta)$, it is proved that $\mathcal{C}(X) \in X_{\lambda\gamma\rho}$.

ii) we have in this case that $BC(U \cup (-U)) = BU$ and therefore:

$$\forall x_1 \in X \exists u_1 \in \rho U : Ax_1 + Bu_1 + D\frac{\delta_1}{\gamma} \in X \ \forall \delta_1 \in \Delta$$

$$Ax_1 + Bu_1 \in \lambda X$$

$$\forall x_2 \in -X \exists u_2 \in -\rho U : Ax_2 + Bu_2 - D\frac{\delta_2}{\gamma} \in X \forall \delta_2 \in \Delta$$

$$Ax_2 + Bu_2 \in -\lambda \widetilde{X}$$

therefore

$$A(\alpha x_1 + \beta x_2) + B(\alpha u_1 + \beta u_2) + D\frac{\mathcal{C}(\Delta \cup (-\Delta))}{\gamma} \in \mathcal{C}(X \cup (-X))$$

$$A(\alpha x_1 + \beta x_2) + B(\alpha u_1 + \beta u_2) \in X$$

$$x_1 \in X, x_2 \in -X, \ \forall \alpha, \beta : \alpha + \beta = 1, \ \alpha \geq 0, \ \beta \geq 0$$

Finally, because $X_0 \subseteq \mathcal{C}(X \cup (-X))$, $\mathcal{C}(X \cup (-X)) \subseteq \Sigma$ and $\Delta \subseteq \mathcal{C}(\Delta \cup (-\Delta))$, it follows that $\mathcal{C}(X \cup (-X)) \in X_{\lambda\gamma\rho}.\square$

Assumption 2.1. The sets X_0, Σ and U are convex, have nonempty interior and the origin belongs to the interior of each of them. Moreover X_0 and Σ are bounded.

We now state a preliminary result, to solve problems P2 and P3.

Let be Σ_0 the maximal p.i.c. set in Σ, which, because of necessary condition of stabilizability for system (1.1), is a nonempty set with the origin in its interior. Necessarily we have $X_0 \subseteq \Sigma_0$, otherwise problem P2 has not solution.

Let us define the following sequences of sets, depending on the parameters $(\lambda, \gamma, \rho) \in \Pi$:

$$\Omega_k(\lambda, \gamma, \rho) = \Sigma_0 \quad k = 0 \tag{2.1}$$

$$\Sigma_k(\lambda, \gamma, \rho) = \Sigma_0 \quad k = 0$$

$$\Omega_k(\lambda, \gamma, \rho) = \{x : \exists u \in \rho U : (Ax + Bu \in \lambda \Sigma_{k-1}(\lambda, \gamma, \rho) \text{ and}$$

$$Ax + Bu + \frac{D\Delta}{\gamma} \subseteq \Sigma_{k-1}(\lambda, \gamma, \rho))\} \quad k > 0$$

$$\Sigma_k(\lambda, \gamma, \rho) = \Omega_k(\lambda, \gamma, \rho) \cap \Sigma_0 \quad k > 0$$

Remark 2.1. similar backward recursions have been used in literature, to solve problems in the context of constrained control and/or optimal control: see e.g. [1], [10], [14], but our technique to compute the sets in the recursion differs from the others used in the above papers because the dependence from the parameters λ, γ and ρ is made explicit in the description of each set (see in this respect the papers [6] and [7]). This fact is of paramount importance in our approach, as you can see in the following section 4.

Lemma 2.1. *If the sets Σ and BU are convex, each of the sets $\Sigma_k(\lambda, \gamma, \rho)$ is convex, $\forall(\lambda, \gamma, \rho) \in \Pi$. If Σ and BU are moreover 0-symmetric, the same is true for each of the sets $\Sigma_k(\lambda, \gamma, \rho)$, $\forall(\lambda, \gamma, \rho) \in \Pi$.*

Proof: The convexity of the sets $\Sigma_k(\lambda, \gamma, \rho)$ can be proved by means of the following chain of implications: Σ convex $\to \Sigma_0$ convex $\to \Omega_1(\lambda, \gamma, \rho)$ convex $\to \Sigma_1$ convex $\to \Omega_2$ convex and so on.

If we apply statement (ii) of theorem 2.1, we have that Σ_0 is 0-symmetric, which implies that Ω_1 is 0-symmetric and so on. $\square$

It is easy to prove that, if for some $\bar{k}$ and for some parameters $(\lambda, \gamma, \rho) \in \Phi$, $X_0 \subseteq \Sigma_{\bar{k}}(\lambda, \gamma, \rho) \subseteq \Omega_{\bar{k}+1}(\lambda, \gamma, \rho)$ then the set $\Sigma_{\bar{k}}(\lambda, \gamma, \rho)$ belongs to the family $X_{\lambda\gamma\rho}$. Moreover if $(\lambda, \gamma, \rho) \in \Phi$, the sequence $\{\Sigma_k(\lambda, \gamma, \rho), \ k = 0, 1, \ldots\}$ converges to the set $\Sigma^*(\lambda, \gamma, \rho)$, otherwise, if (λ, γ, ρ) does not belong to the feasible parameter set Φ, we have that, for some k, the set $\Sigma_k(\lambda, \gamma, \rho)$ becomes empty and/or X_0 is not a subset of $\Sigma_k(\lambda, \gamma, \rho)$.

If Σ_0 is a polytope and U is a polyhedron, we are able to explicitly describe the sets defined in (2.1), as we can see in the following theorem 2.2. In the sequel we'll show that it is possible to remove the above assumption on Σ_0, because, in our framework, the fact that Σ is a polytope implies that Σ_0 is a polytope, too (see remark 3.1).

Theorem 2.2. *If $\Sigma_0 = \{x : Gx \leq v\}$ and $U = \{u : Fu \leq w\}$, the sets $\Sigma_k(\lambda, \gamma, \rho)$ and $\Omega_k(\lambda, \gamma, \rho)$, if nonempty, are polyhedra, for all values of λ, γ, ρ, and have the following expression:*

$$\Omega_k(\lambda, \gamma, \rho) = \{x : \tilde{G}(k)x \leq \tilde{v}_{\lambda\gamma\rho}(k)\}$$

$$\tilde{G}(k) = (Q^{(1)}(k-1) + Q^{(2)}(k-1))\hat{G}(k-1)A$$

$$\tilde{v}_{\lambda\gamma\rho}(k) = (Q^{(1)}(k-1) + \lambda Q^{(2)}(k-1))\hat{v}_{\lambda\gamma\rho}(k-1) - Q^{(1)}(k-1)\frac{z(k-1)}{\gamma} \\ + \rho Q^{(3)}(k-1)w$$

$$\Sigma_k(\lambda, \gamma, \rho) = \{x : \hat{G}(k)x \leq \hat{v}_{\lambda\gamma\rho}(k)\}$$

where

$$\hat{G}(k) = \begin{pmatrix} \tilde{G}(k) \\ G \end{pmatrix} \quad \hat{v}_{\lambda\gamma\rho}(k) = \begin{pmatrix} \tilde{v}_{\lambda\gamma\rho}(k) \\ v \end{pmatrix}$$

$$\hat{G}(0) = G \quad \hat{v}(0) = v$$

the matrices $Q^{(1)}(k), Q^{(2)}(k)$ and $Q^{(3)}(k)$ are suitable column partitions of the matrix $Q(k)$ defined as:

$$\{\text{rows of } Q(k)\} = \left\{ \text{generators of } \mathcal{R} \begin{pmatrix} \hat{G}(k)B \\ \hat{G}(k)B \\ F \end{pmatrix}^\perp \cap P \right\}$$

$$Q(k) = (Q^{(1)}(k)|Q^{(2)}(k)|Q^{(3)}(k))$$

and the ith component of vector $z(k)$ is defined as

$$[z(k)]_i = \max_{\delta \in \Delta} [\hat{G}(k)]_i D\delta$$

where $[\hat{G}(k)]_i$ denotes the ith row of the matrix $\hat{G}(k)$.

Proof: the set $\Omega_1(\lambda, \gamma, \rho)$ is the set of all the states starting from which at some time T we can reach the set $\lambda \Sigma_0$ at time $T+1$, with some input in the set ρU, and from which, with the same input, we can reach the set Σ_0, in one step of time, for all values of the disturbance, belonging to the set $\frac{\Delta}{\gamma}$. The set $\Sigma_1(\lambda, \gamma, \rho)$ is the set of states in Σ_0 with the above properties, and so on. Obviously the choice of instant T is irrilevant, because we are in a time invariant context. With these premises, we can say that at time $T+1$ we must have:

$$\begin{cases} Gx(T+1) \leq \lambda v & \Delta = \{0\} \\ Gx(T+1) \leq v & \Delta \neq \{0\} \end{cases}$$

Substituting for $x(T+1)$ we have:

$$\begin{cases} G(Ax(T) + Bu(T)) \leq \lambda v \\ G(Ax(T) + Bu(T) + D\delta(T)) \leq v \quad \forall \delta(T) \in \frac{\Delta}{\gamma} \end{cases}$$

These last inequalities can be rewritten as:

$$\begin{cases} GBu(T)) \leq \lambda v - GAx(T) \\ GBu(T) \leq v - GAx(T) - \frac{z(0)}{\gamma} \end{cases}$$

where each component of the vector $z(0)$ is defined by:

$$[z(0)]_i = \max_{\delta \in \Delta} G_i D\delta$$

Because the input $u(T)$ must belong to the set ρU we have:

$$\begin{pmatrix} GB \\ GB \\ F \end{pmatrix} u(T) \leq \begin{pmatrix} \lambda v - GAx(T) \\ v - GAx(T) - \frac{z(0)}{\gamma} \\ \rho w \end{pmatrix} \tag{2.2}$$

In view of the dual nonvoidness condition (theorem 1.1), the set of all bounds that make the latter equation feasible is given by:

$$Q(0) \begin{pmatrix} \lambda v - GAx(T) \\ v - GAx(T) - \frac{z(0)}{\gamma} \\ \rho w \end{pmatrix} \geq 0$$

where $Q(0)$ is the matrix, whose rows are the generators of the pointed polyhedral cone $\mathcal{R} \begin{pmatrix} GB \\ GB \\ F \end{pmatrix}^{\perp} \cap P$.

Therefore, partitioning the matrix $Q(0)$ in accord to the dimensions of the row blocks in the vector $\begin{pmatrix} \lambda v - GAx(T) \\ v - GAx(T) - \frac{z(0)}{\gamma} \\ \rho w \end{pmatrix}$, as $Q(0) = (Q^{(1)}(0)|Q^{(2)}(0)|Q^{(3)}(0))$ we have:

$$\tilde{G}(1)x(T) \leq \tilde{v}_{\lambda\gamma\rho}(1) \tag{2.3}$$

where

$$\tilde{G}(1) = (Q^{(1)}(0) + Q^{(2)}(0))GA$$

$$\tilde{v}_{\lambda\gamma\rho}(1) = (\lambda Q^{(1)}(0) + Q^{(2)}(0))v - Q^{(2)}(0)\frac{z(0)}{\gamma} + \rho Q^{(3)}(0)w$$

The inequality (2.3) defines the set of states $\Omega_1(\lambda, \gamma, \rho)$, showing at the same time that it is a polyhedron. The set $\Sigma_1(\lambda, \gamma, \rho)$ is obtained intersecting $\Omega_1(\lambda, \gamma, \rho)$ with the set Σ_0 and hence:

$$\Sigma_1(\lambda, \gamma, \rho) = \{x : \hat{G}(1)x \leq \hat{v}_{\lambda\gamma\rho}(1)\}$$

where

$$\hat{G}(1) = \begin{pmatrix} \tilde{G}(1) \\ G \end{pmatrix} \qquad \hat{v}_{\lambda\gamma\rho}(1) = \begin{pmatrix} \tilde{v}_{\lambda\gamma\rho}(1) \\ v \end{pmatrix}$$

At this point it is easy to see that generalizing the above formulas to the generic instant of time $t = T - k$, the expressions of $\Omega_k(\lambda, \gamma, \rho)$ and of $\Sigma_k(\lambda, \gamma, \rho)$ are obtained. $\square$

We can state the following corollary:

Corollary 2.1. *If $\Delta = \{0\}$, each of the sets $\Sigma_k(\lambda, \gamma, \rho)$ is nonempty and has the origin in its interior, for all values of the parameters λ, ρ.*

Proof: Because the bound vector $\hat{v}_{\lambda\gamma\rho}(k)$ is strictly positive, $\forall \lambda >, \forall \rho > 0, \forall k$ the desired conclusion follows. $\square$

3. Main results

Because in the description of the sets in the statement of theorem 2.2 the bounds explicitly depend on the parameters, no recomputation of the sequences is required for a different choice of the parameters. This fact gives us the possibility of defining some sequences of parameters, which are the main tools to solve a problem in the class (P3).

Let us consider a triple $(\lambda_0, \gamma_0, \rho_0)$, $\lambda_0 \geq 0, \gamma_0 \geq 0, \rho_0 \geq 0$ and the set $\Sigma^*(\lambda_0, \gamma_0, \rho_0)$. Initialize the sequences (2.1) as follows:

$$\Sigma_0(\lambda, \gamma, \rho) = \Sigma^*(\lambda_0, \gamma_0, \rho_0)$$

$$\Omega_0(\lambda, \gamma, \rho) = \Sigma^*(\lambda_0, \gamma_0, \rho_0)$$

and define the sequences:

$$\{\bar{\lambda}_k, \quad k = 0, 1, \ldots\}$$
$$\{\bar{\gamma}_k, \quad k = 0, 1, \ldots\} \tag{3.1}$$
$$\{\bar{\rho}_k, \quad k = 0, 1, \ldots\}$$

where $\bar{\lambda}_k$ is the minimum value of $\lambda > 0$ such that $X_0 \subseteq \Sigma_k(\lambda, \gamma_0, \rho_0) \subseteq \Omega_{k+1}(\lambda, \gamma_0, \rho_0), \lambda_0 = \infty$, $\bar{\gamma}_k$ is the minimum value of $\gamma > 0$ such that $X_0 \subseteq \Sigma_k(\lambda_0, \gamma, \rho_0) \subseteq \Omega_{k+1}(\lambda_0, \gamma, \rho_0)\gamma_0 = \infty$, $\bar{\rho}_k$ is the minimum value of $\rho > 0$ such that $X_0 \subseteq \Sigma_k(\lambda_0, \gamma_0, \rho) \subseteq \Omega_{k+1}(\lambda_0, \gamma_0, \rho), \rho_0 = \infty$.

We can state the following

Lemma 3.1. *The sequences* $\{\bar{\lambda}_k, \quad k = 0, 1, \ldots\}$, $\{\bar{\gamma}_k, \quad k = 0, 1, \ldots\}$, $\{\bar{\rho}_k, \quad k = 0, 1, \ldots\}$ *are nonincreasing.*

Proof: By definition $\bar{\lambda}_{k+1}$ is the minimum value of λ such that $X_0 \subseteq \Sigma_{k+1}(\lambda, \gamma_0, \rho_0) \subseteq \Omega_{k+2}(\lambda, \gamma_0, \rho_0)$ and $\bar{\lambda}_k$ is the minimum value of λ such that $X_0 \subseteq \Sigma_k(\lambda, \gamma_0, \rho_0) \subseteq \Omega_{k+1}(\lambda, \gamma_0, \rho_0)$. Therefore, if $\bar{\lambda}_{k+1} > \bar{\lambda}_k$, we have that $\Sigma_k(\bar{\lambda}_{k+1}, \gamma_0, \rho_0) \subseteq \Omega_{k+1}(\bar{\lambda}_{k+1}, \gamma_0, \rho_0)$, which implies that $\Sigma_{k+1}(\bar{\lambda}_{k+1}, \gamma_0, \rho_0) = \Sigma_k(\bar{\lambda}_{k+1}, \gamma_0, \rho_0)$ and $\Omega_{k+2}(\bar{\lambda}_{k+1}, \gamma_0, \rho_0) = \Omega_{k+1}(\bar{\lambda}_{k+1}, \gamma_0, \rho_0)$. This means that necessarily $\bar{\lambda}_{k+1} \leq \bar{\lambda}_k$.

Similar argumentations apply to the other sequences. $\square$

Remark 3.1. With the above lemma 3.1 we can prove that if system (1.1) is stabilizable and Σ is a convex polytope, then the set Σ_0 is a convex polytope. To do this, let us initialize the sequences (2.1) as follows:

$$\gamma_0 = \infty$$
$$\rho_0 = \infty$$
$$\Sigma_0(\lambda, \gamma_0, \rho_0) = \Sigma$$
$$\Omega_0(\lambda, \gamma_0, \rho_0) = \Sigma$$

and let us assume that $\bar{\lambda}_0 > 1$ (otherwise we trivially have $\Sigma_0 = \Sigma$). Because the system is stabilizable we have $\lim_{k \to \infty} \{\bar{\lambda}_k\} < 1$, because a compact contractive controllable set surely exists (see e.g. [2]). Hence, because the sequence $\{\bar{\lambda}_k, k = 0, 1, \ldots\}$ is nonincreasing, there exists some value $k = \bar{k}$ such that $\Sigma^*(1, \gamma_0, \rho_0) = \Sigma_{\bar{k}}(1, \gamma_0, \rho_0) = \Sigma_0$. This means that Σ_0 is described by a finite set of linear inequalities and therefore it is a polytope.

Given $(\lambda, \gamma, \rho) \in \Phi$, the backward recursion defined in (2.1) converges to the convex set $\Sigma^*(\lambda, \gamma, \rho)$. This set, obviously, might be not polyhedral. The following theorem 3.1, generalizing a result stated in [1], show that, if we relax our requirements, a polyhedral approximation of the set $\Sigma^*(\lambda, \gamma, \rho)$ can be found.

Theorem 3.1. *Assume that, given the parameters $(\lambda, \gamma, \rho) \in \Phi$, the set $\Sigma^*(\lambda, \gamma, \rho)$ is a convex compact set containing the origin in its interior. Then for every λ', γ', ρ' such that $1 \geq \lambda' > \lambda$, $\gamma' > \gamma$, $\rho' > \rho$ there exists a k such that $\Sigma_k(\lambda, \gamma, \rho) \in X_{\lambda' \gamma' \rho'}$.*

Proof: The set $\Sigma^*(\lambda, \gamma, \rho)$ will be denoted in this proof by the symbol Σ^* for simplicity. Let us make the following positions:

$$\mu_\lambda = \frac{\lambda'}{\lambda}$$

$$\mu_\rho = \max \mu : \forall x \in \mu \Sigma^* \; \exists u \in \rho' U : Ax + Bu \in \lambda \mu \Sigma^*$$

$$\mu_\gamma = \max \mu \leq \mu_\rho : \forall x \in \mu \Sigma^* \; \exists u \in \rho' U : Ax + Bu + \frac{D\Delta}{\gamma'} \subseteq \Sigma^*$$

It is easy to see that $\mu_\lambda > 1$, $\mu_\rho > 1$ and $\mu_\gamma > 1$. In fact $\mu_\lambda > 1$ by definition of λ', $\mu_\rho > 1$ because if the set Σ^* is λ- contractive controllable, with the constraint $u(t) \in \rho U \; \forall t$, being $\alpha = \frac{\rho'}{\rho} > 1$, we can write:

$$\forall x \in \Sigma^* \; \exists u \in \rho U : \alpha Ax + \alpha Bu \in \lambda \alpha \Sigma^*$$

which implies that

$$\forall x \in \alpha \Sigma^* \; \exists u \in \rho' U : Ax + Bu \in \lambda \alpha \Sigma^*$$

Therefore $\mu_\rho \geq \alpha$ and hence $\mu_\rho > 1$.
Finally, because $\Sigma^* \subseteq X_{\lambda \gamma \rho}$, a set $\bar{\Sigma} \in \lambda \Sigma^*$ exists, such that

$$\exists u \in \rho U : Ax + Bu \in \bar{\Sigma}$$

$$\bar{\Sigma} + D\frac{\Delta}{\gamma} \subseteq \Sigma^*$$

Therefore some $\tilde{\lambda} < 1$ depending on $\frac{\gamma'}{\gamma}$ exists such that $\bar{\Sigma} + D\frac{\Delta}{\gamma'} \subseteq \tilde{\lambda} \Sigma^*$. Taking some $\alpha = \frac{1}{\lambda} > 1, \alpha < \mu_\rho$, we have that the set $\alpha \bar{\Sigma}$ is such that

$$\forall x \in \Sigma^* \ \exists u \in \rho U : Ax + Bu \in \alpha \bar{\Sigma} \ \text{ and } \ \alpha \bar{\Sigma} + \alpha D \frac{\Delta}{\gamma'} \subseteq \Sigma^*$$

This means that

$$\forall x \in \alpha \Sigma^* \ \exists u \in \rho U : Ax + Bu + \alpha D \frac{\Delta}{\gamma'} \subseteq \Sigma^*$$

Therefore

$$\forall x \in \alpha \Sigma^* \ \exists u \in \rho U : Ax + Bu + D \frac{\Delta}{\gamma'} \subseteq \Sigma^*$$

and finally, because $\mu_\gamma \geq \alpha$ necessarily we have $\mu_\gamma > 1$.

Let us make the position $\bar{\mu} = \min\{\mu_\lambda, \mu_\rho, \mu_\gamma\}$. If Σ^* is a convex compact set with the origin in its interior, some k exists such that $\Sigma^* \subseteq \Sigma_k(\lambda, \gamma, \rho) \subseteq \bar{\mu}\Sigma^*$. The set $\bar{\mu}\Sigma^*$ is λ- contractive controllable with the constraint $u(t) \in \rho'U, \forall t$, and hence

$$\forall x \in \Sigma_k(\lambda, \gamma, \rho) \ \exists u \in \rho'U :$$
$$Ax + Bu \in \bar{\mu}\lambda\Sigma^* \subseteq \mu_\lambda \lambda \Sigma^* = \lambda'\Sigma^* \subseteq \lambda'\Sigma_k(\lambda, \gamma, \rho)$$

Moreover $\forall x \in \bar{\mu}\Sigma^* \ \exists u \in \rho U : Ax + Bu + D\frac{\Delta}{\gamma'} \in \Sigma^*$, which implies that

$$\forall x \in \Sigma_k(\lambda, \gamma, \rho) \ \exists u \in \rho U :$$
$$Ax + Bu + D\frac{\Delta}{\gamma'} \in \Sigma^* \subseteq \Sigma_k(\lambda, \gamma, \rho)$$

If $X_0 \subseteq \Sigma^*(\lambda, \gamma, \rho)$, $X_0 \subseteq \Sigma_k(\lambda, \gamma, \rho)$ and therefore we can conclude that $\Sigma_k(\lambda, \gamma, \rho) \in X_{\lambda', \gamma', \rho'}$. $\square$

We give now a simple lemma, which will be useful in the sequel. Let us consider two nonempty convex polyhedra P_1 and P_2 described respectively by the inequalities $Gx \leq v$ and $Fx \leq w$.

Lemma 3.2. *The system $Gx \leq v$, $G \in R^{m \times n}$, is satisfied by any point of P_2 if and only if*

$$\max_{x \in P_2} G_i x \leq v_i \ \ i = 1 \dots m \tag{3.2}$$

where G_i denotes the ith row vector of the matrix G.

Proof: The proof is trivial and therefore it is omitted.$\square$

Let us now study the problem of the computation of the sequences defined in (3.1). If we want to evaluate $\bar{\lambda}_k$, we have to solve the problem:

$$\min \lambda, \ \text{ subject to}$$

$$\tilde{G}(k+1)x \leq \tilde{v}_{\lambda \gamma_0 \rho_0}(k+1) \ \forall x \in \Sigma_k(\lambda, \gamma_0, \rho_0)$$

$$\hat{G}(k)x \leq \hat{v}_{\lambda \gamma_0 \rho_0}(k) \ \forall x \in X_0$$

Applying lemma 3.2, this problem is equivalent to the problem:

$$\min \lambda, \quad \text{subject to}$$

$$\tilde{v}_{\lambda \gamma_0 \rho_0}(k+1) \geq z(\lambda)$$

$$\hat{v}_{\lambda \gamma_0 \rho_0}(k) \geq s$$

where, denoting by M_i the ith row vector of some matrix M, the ith component of the vector $z(\lambda)$ is:

$$z_i(\lambda) = \max_{x \in \Sigma_k(\lambda, \gamma_0, \rho_0)} (\tilde{G}(k+1))_i x$$

and the jth component of the vector s is:

$$s_j = \max_{x \in X_0} (\hat{G}(k))_j x$$

Because using the approach recalled in equation (1.2) we are able to explicitly express each of the maxima in above definition of $z(\lambda)$ as a function of the parameter λ, which appears only in the bound vector of the inequality defining the set $\Sigma_k(\lambda, \gamma_0, \rho_0)$, the problem reduces to the minimization the value of the scalar parameter λ, subject to a set of nonlinear inequalities, which defines an open segment, bounded below. The same technique applies also to the sequences $\{\bar{\gamma}_k\}$ and $\{\bar{\rho}_k\}$.

It is clear that the number of linear inequalities which describe the sets $\Sigma_k(\lambda, \gamma, \rho)$ and $\Omega_k(\lambda, \gamma, \rho)$ increases as k increases, and also the numerical effort required to compute the sequences defined in (3.1) increases, step by step. Therefore it is important to have an efficient algorithm which detects and eliminates all redundant inequalities, step by step. This task is hard if the bound vector of the inequalities depends on a parameter, as in our case. Some research effort is in progress on these computational problems.

At this point we can state the following

Theorem 3.2. *For all* $\begin{pmatrix} \lambda \\ \gamma \\ \rho \end{pmatrix} \in \Phi$, *for all* $\begin{pmatrix} \lambda' \\ \gamma' \\ \rho' \end{pmatrix} \in \Phi$, *such that*

$\begin{pmatrix} \lambda' \\ \gamma' \\ \rho' \end{pmatrix} > \begin{pmatrix} \lambda \\ \gamma \\ \rho \end{pmatrix}$, *the set* $\Sigma^*(\lambda', \gamma', \rho')$ *is nonempty and it is polyhedral.*

Proof: The proof is constructive. In fact we show that an algorithm exists which converges to the set $\Sigma^*(\lambda', \gamma', \rho')$ in a finite number of steps. The algorithm is the following:

Algorithm 3.1.

STEP 1: Initialize $\lambda_0 = 1$, $\gamma_0 = \infty$, $\rho_0 = \infty$, $h = 0$, $k = 0$.

STEP 2: Compute $\Sigma^*(\lambda_0, \gamma_0, \rho_0)$. As shown in remark 3.1, this last set is polyhedral, and it can be computed in a finite number of steps, say h_0. Set $h = h + h_0$.

STEP 3: Compute $\bar{\lambda}_k$, increasing k until $\bar{\lambda}_k = \lambda'$, $k = \bar{k}$. The sequence $\{\bar{\lambda}_k\}$ converges to a value less or equal to λ, as k goes to ∞. In fact, if $\bar{\lambda}_\infty = \lim_{k\to\infty}\{\bar{\lambda}_k\}$ we have that, given some $\bar{\lambda}$, $\Sigma^*(\bar{\lambda},\infty,\infty)$ is empty if $\bar{\lambda} < \bar{\lambda}_\infty$ and it is nonempty if $\bar{\lambda} \geq \bar{\lambda}_\infty$. Moreover, because $(\lambda,\gamma,\rho) \in \Phi$, the set $\Sigma^*(\lambda,\gamma,\rho)$ is nonempty, and it is a subset of $\Sigma^*(\lambda,\infty,\infty)$. Therefore $\bar{\lambda}_\infty \leq \lambda$. Because the sequence is nonincreasing (lemma 3.1), the above value $\bar{k}$ exists and it is finite. Set $h = h + \bar{k}$, $k = 0$, $\lambda_0 = \lambda'$. $\Sigma^*(\lambda_0,\gamma_0,\rho_0) = \Sigma_{\bar{k}}(\lambda_0,\gamma_0,\rho_0)$.

STEP 4: Compute $\bar{\gamma}_k$, increasing k until $\bar{\gamma}_k = \gamma'$, $k = \bar{k}$. The sequence $\{\bar{\gamma}_k\}$ converges to a value less or equal to γ', and because the sequence is nonincreasing (lemma 3.1), the above value $\bar{k}$ exists and it is finite. Set $h = h + \bar{k}$, $k = 0$, $\gamma_0 = \gamma'$. $\Sigma^*(\lambda_0,\gamma_0,\rho_0) = \Sigma_{\bar{k}}(\lambda_0,\gamma_0,\rho_0)$.

STEP 5: Compute $\bar{\rho}_k$, increasing k until $\bar{\rho}_k = \rho'$, $k = \bar{k}$. The sequence $\{\bar{\rho}_k\}$ converges to a value less or equal to ρ', and because the sequence is nonincreasing (lemma 3.1), the above value $\bar{k}$ exists and it is finite. Set $h = h + \bar{k}$, $\rho_0 = \rho'$. $\Sigma^*(\lambda_0,\gamma_0,\rho_0) = \Sigma_{\bar{k}}(\lambda_0,\gamma_0,\rho_0)$.

STEP 6: The required set $\Sigma^*(\lambda',\gamma',\rho')$ is equal to $\Sigma_{\bar{k}}(\lambda_0,\gamma_0,\rho_0)$, which is nonempty, and, because it has been computed in h steps of the recursion (2.1), it is a polyhedron. STOP. $\square$

Using the same symbols defined in theorem 2.2, we have:

Theorem 3.3. *Let be* $\begin{pmatrix} \lambda \\ \gamma \\ \rho \end{pmatrix} \in \Phi$ *a triple such that the set* $\Sigma^*(\lambda,\gamma,\rho)$ *is polyhedral. The set of all and nothing but the inputs that solve problem P2 with this choice of parameters is defined by:*

$$\begin{pmatrix} \hat{G}(k)B \\ \hat{G}(k)B \\ F \end{pmatrix} u(t) \leq \begin{pmatrix} \lambda\mu(x(t))\hat{v}(k) - \hat{G}(k)Ax(t) \\ \hat{v}(k) - \hat{G}(k)Ax(t) - \frac{z(k)}{\gamma} \\ \rho w \end{pmatrix} \tag{3.3}$$

where k is such that $\Sigma_k(\lambda,\rho,\gamma) = \Sigma^*(\lambda,\gamma,\rho) = \{x : \hat{G}(k)x \leq \hat{v}(k)\}$, and $\mu(x(t)) = \inf \alpha : \hat{G}(k)x(t) \leq \alpha\hat{v}(k)$.

Proof: The expression of the set of inputs can be easily deduced, generalizing the inequality (2.2) to generic k. $\square$

Theorem 5 above shows that it is possible to compute a triple (λ',γ',ρ') arbitrarily close to any point in the efficient subset of Φ, where, as it is well known, the efficient subset $\mathcal{E}$ is the set of all optimal points in Paretian sense, i.e.

$$\mathcal{E} = \left\{ \begin{pmatrix} \lambda_e \\ \gamma_e \\ \rho_e \end{pmatrix} \in \varPhi : \nexists \begin{pmatrix} \lambda \\ \gamma \\ \rho \end{pmatrix} \in \varPhi, \right.$$

$$\left. \begin{pmatrix} \lambda \\ \gamma \\ \rho \end{pmatrix} \neq \begin{pmatrix} \lambda_e \\ \gamma_e \\ \rho_e \end{pmatrix} : \begin{pmatrix} \lambda_e \\ \gamma_e \\ \rho_e \end{pmatrix} - \begin{pmatrix} \lambda \\ \gamma \\ \rho \end{pmatrix} \geq 0 \right\}$$

Because to each triple $(\lambda', \gamma', \rho')$ is associated a polytope $\Sigma^*(\lambda', \gamma', \rho')$, and because this last polytope is the maximal set in the family $X_{\lambda \gamma \rho}$, Theorem 6 gives a description of all and nothing but the control laws such that performance $(\lambda', \gamma', \rho')$ is assured.

Notice moreover that in the above algorithm 3.1 steps 3, 4 and 5 may be performed in any order. The resulting set $\Sigma^*(\lambda', \gamma', \rho')$ is obviously the same, for all order. The number of steps h on the contrary may in general depend on the order itself.

This last remark gives us the tool to find a near optimal solution of each problem in the class P3, as we can see, examining again the example shortly outlined in section 2. The problem is: given a stabilizable linear dynamic system and given the sets Σ, X_0 and the parameter $\rho = \bar{\rho}$, solve problem P2, assuring some rate of exponential convergence to the origin, and finally minimizing the parameter γ. From algorithm 3.1, we can derive the following algorithm:

Algorithm 3.2.

STEP 1: Initialize $\lambda_0 = 1$, $\gamma_0 = \infty$, $\rho_0 = \infty$, $h = 0$, $k = 0$.

STEP 2: Compute $\Sigma^*(\lambda_0, \gamma_0, \rho_0)$. As shown in remark 3.1, this last set is polyhedral, and it can be computed in a finite number of steps, say h_0. Set $h = h + h_0$.

STEP 3: Compute $\bar{\rho}_k$, increasing k until $\bar{\rho}_k = \bar{\rho}$, $k = \bar{k}$, or $k > k_{max}$. If $\bar{\rho}_{\bar{k}} = \bar{\rho}$, set $\rho_0 = \bar{\rho}, h = h + \bar{k}, k = 0$, compute $\Sigma_{\bar{k}}(\lambda_0, \gamma_0, \rho_0)$ and go to STEP 4, otherwise STOP. The value k_{max} is the maximum admissible number of iterations, and it takes account of dimension of the problem and of computational limitations.

STEP 4: Compute $\bar{\lambda}_k$, increasing k until a satisfactory level $\bar{\lambda}_k = \lambda'$, $k = \bar{k}$, is obtained. Set $h = h + \bar{k}$, $k = 0$, $\lambda_0 = \lambda'$. $\Sigma^*(\lambda_0, \gamma_0, \rho_0) = \Sigma_{\bar{k}}(\lambda_0, \gamma_0, \rho_0)$.

STEP 5: Compute $\bar{\gamma}_k$, increasing k until a satisfactory level $\bar{\gamma}_k = \gamma'$, $k = \bar{k}$, is obtained. Set $h = h + \bar{k}$, $\gamma_0 = \gamma'$. $\Sigma^*(\lambda_0, \gamma_0, \rho_0) = \Sigma_{\bar{k}}(\lambda_0, \gamma_0, \rho_0)$.

STEP 6: Given the polyhedral set $\Sigma^*(\lambda_0, \gamma_0, \rho_0) = \Sigma^*(\lambda', \gamma', \bar{\rho})$, computed at step 5, apply theorem 3.3, and find a control law. STOP

With respect to the above algorithm we remark that only the value $\bar{\rho}$ is a priori known, because it represents hard constraints on the input variables. Notice moreover that, because the minimum attainable value of $\bar{\lambda}_k$ at step

4 depends on $\bar{\rho}$, and because the minimum attainable value of $\bar{\gamma}_k$ at step 5 depends on $\bar{\rho}$ and on λ', obtained at step 4, it is clear that the order in which we perform the minimization of the parameters is relevant, with repsect to the emphasis we want give to each parameter.

If the above algorithm fails, i.e. if it ends at STEP 3, two cases are possible: $(1, \infty, \bar{\rho}) \notin \Phi$ or the set $\Sigma^*(1, \infty, \bar{\rho})$ cannot be computed in a number of iterations, compatible with our computation capability. In both cases we have to reformulate a new problem, with a smaller set of initial states and/or with a larger value of $\bar{\rho}$. A reinitialization of the algorithm is not required: in fact, because we have computed the sequence $\{\bar{\rho}_k, \; k = 0, 1, \ldots k_{max}\}$, we can assume one of the elements $\bar{\rho}_{\tilde{k}}$ of the sequence as a new feasible value of $\bar{\rho}$, and the set $\Sigma_{\tilde{k}}(1, \infty, \bar{\rho})$ can be immediately derived. Similarly, given some $\tilde{k} < k_{max}$, we can compute a subset of X_0, contained in $\Sigma_{\tilde{k}}(1, \infty, \bar{\rho})$.

To end this section, we give a simple numerical example, where we explicitly compute the sequence $\{\bar{\lambda}_k\}$:

EXAMPLE 1:
Let the set X_0 be a neighbouround of the origin, described by the inequalities:

$$\begin{pmatrix} 1 & 0 \\ 0 & 1 \\ -1 & 0 \\ 0 & -1 \end{pmatrix} x \le \epsilon \begin{pmatrix} 1 \\ 1 \\ 1 \\ 1 \end{pmatrix} \qquad 0 < \epsilon \le 0.5$$

Let us consider the set Σ described by the inequalities:

$$\begin{pmatrix} 1 & 1 \\ -1 & 0 \\ 0 & -1 \end{pmatrix} x \le \begin{pmatrix} 1 \\ 0.5 \\ 1 \end{pmatrix}$$

and the system described by

$$x(t+1) = Ax(t) + Bu(t)$$

where

$$A = \begin{pmatrix} 1 & 2 \\ -1 & -1 \end{pmatrix} \qquad B = \begin{pmatrix} 1 \\ 0 \end{pmatrix}$$

We make the positions $\rho_0 = \infty$, $\gamma_0 = \infty$, $\Sigma_0(\lambda, \gamma_0, \rho_0) = \Sigma$ and we want to compute the sequence $\bar{\lambda}_k$. We obtain:

$$\Omega_1(\lambda, \gamma_0, \rho_0) = \left\{ x : \begin{pmatrix} 1 & 1 \\ -1 & -1 \end{pmatrix} x \le \lambda \begin{pmatrix} 1 \\ 1.5 \end{pmatrix} \right\}$$

$$\bar{\lambda}_0 = 1$$

$$\Sigma_1(\lambda, \gamma_0, \rho_0) = \Sigma \cap \Omega_1(\lambda, \gamma_0, \rho_0)$$

$$\Omega_2(\lambda, \gamma_0, \rho_0) = \left\{ x : \begin{pmatrix} 1 & 1 \\ -1 & -1 \end{pmatrix} x \le \lambda \begin{pmatrix} 1 \\ 0.5 + \lambda \end{pmatrix} \right\}$$

The value $\bar{\lambda}_1$ is the minimum λ such that the following inequalities are fulfilled:

$$\max_{x \in \Sigma_1(\lambda, \gamma_0, \rho_0)} (1\ 1)x \le \lambda$$

$$\max_{x \in \Sigma_1(\lambda, \gamma_0, \rho_0)} (-1\ -1)x \le \lambda(0.5 + \lambda)$$

$$\max_{x \in X_0}(1\ 1)x \le \lambda$$

$$\max_{x \in X_0}(-1\ -1)x \le 1.5\lambda$$

Appling 2.2 we obtain:

$$\max_{x \in \Sigma_1(\lambda, \gamma_0, \rho_0)} (1\ 1)x = \lambda$$

$$\max_{x \in \Sigma_1(\lambda, \gamma_0, \rho_0)} (-1\ -1)x = \min\{1.5; 1.5\lambda\} = 1.5\lambda$$

$$\max_{x \in X_0}(1\ 1)x = 2\epsilon$$

$$\max_{x \in X_0}(-1\ -1)x = 2\epsilon$$

and therefore

$$\bar{\lambda}_1 = 1$$

We perform one more step:

$$\Sigma_2(\lambda, \gamma_0, \rho_0) = \Sigma \cap \Omega_2(\lambda, \gamma_0, \rho_0)$$

$$\Omega_3(\lambda, \gamma_0, \rho_0) = \Omega_2(\lambda, \gamma_0, \rho_0)$$

The value $\bar{\lambda}_2$ is the minimum λ such that the following inequalities are fulfilled:

$$\max_{x \in \Sigma_2(\lambda, \gamma_0, \rho_0)} (1\ 1)x \le \lambda$$

$$\max_{x \in \Sigma_2(\lambda, \gamma_0, \rho_0)} (-1\ -1)x \le \lambda(0.5 + \lambda)$$

$$\max_{x \in X_0}(1\ 1)x \le \lambda$$

$$\max_{x \in X_0}(-1\ -1)x \le \lambda(0.5 + \lambda)$$

Applying again (1.2):

$$\max_{x \in \Sigma_2(\lambda, \gamma_0, \rho_0)} (1\ 1)x = \lambda$$

$$\max_{x \in \Sigma_2(\lambda, \gamma_0, \rho_0)} (-1\ -1)x = \lambda(0.5 + \lambda)$$

$$\max_{x \in X_0}(1\ 1)x = 2\epsilon$$

$$\max_{x \in X_0}(-1\ -1)x = 2\epsilon$$

and therefore

$$\bar{\lambda}_2 = \max\{2\epsilon,\ \sqrt{0.0625 + 2\epsilon} - 0.25\}$$

Finally, because $\Omega_3(\lambda, \gamma_0, \rho_0) = \Omega_2(\lambda, \gamma_0, \rho_0)$, we have also that $\Sigma_3(\lambda, \gamma_0, \rho_0) = \Sigma_2(\lambda, \gamma_0, \rho_0)$, and therefore $\bar{\lambda}_k = \bar{\lambda}_2, \forall k > 2$.

4. Some notes on the robustness

In this section we outline the case in which the matrix A is not exactly known. More specifically we address two different problems. In the first one we give conditions under which a control law of no preassigned structure exists such that a known set in the family $X_{\lambda\gamma\rho}$ remains in the family for all matrices A in a polytope $\mathcal{A}$ of a given shape. If the class of such controls is nonempty, we select the subclass such that the radius of the polytope $\mathcal{A}$ is maximized.

In the second problem, we restrict our attention to the class of linear static feedback control laws. If problem P2 (or P3) has a solution with this restriction, we are able to handle the general case in which the matrix A depends on some parameters vector.

Let us now examine the first problem. We have

$$A \in \mathcal{A}(\eta) \tag{4.1}$$

where $\mathcal{A}(\eta)$ is a polytope of matrices in the space $R^{n \times n}$, described as:

$$\mathcal{A}(\eta) = A_0 + \eta\mathcal{C}\{A^{(i)}\}, \quad i = 1 \ldots p \tag{4.2}$$

where A_0 is the nominal matrix of the system, $\eta \in R_+$ is a parameter to be maximized and, for a fixed η, $\eta A^{(i)}, i = 1 \ldots p$ are the vertices of the polytope $\mathcal{A}(\eta) - A_0$.

Let us suppose we have solved some control problem P2 or P3 for the nominal system, i.e. we have found a triple (λ, γ, ρ) and the polytope $\Sigma^*(\lambda, \gamma, \rho) = \{x : Mx \le s\}$, which will be denoted shortly Σ^*.

We want to evaluate the maximal value η^* of the radius η such that the set Σ^* is a solution also for the uncertain system, i.e. it enjoys the properties i) and ii) in proposition 1.1, for all $A \in \mathcal{A}(\eta)$. We want moreover to characterize the corresponding robust control.

The solution is given in the following:

Theorem 4.1. *The set Σ^* enjoys requirements i) and ii) of proposition 1.1 $\forall A \in \mathcal{A}(\eta)$ if and only if $\Sigma^* \subseteq \Omega(\eta)$, where $\Omega(\eta) = \{x : \exists u \in \rho U : ((A_0 + \eta A_i)x + Bu \in \lambda\Sigma^*$ and $(A_0 + \eta A_i)x + Bu + \frac{D\Delta}{\gamma} \subseteq \Sigma^*)\ i = 1 \ldots p\}$.*

Proof: the proof is rather trivial and it is omitted for shortness. $\square$

Using the same technique of the proof of theorem 2.2, we are able to explicitly compute the set $\Omega(\eta)$, which is a polyhedron described by $\{x : \tilde{M}(\eta)x \leq \tilde{s}\}$, where

$$
\tilde{M}(\eta) = Q \begin{pmatrix} M(A_0 + \eta A^{(1)}) \\ M(A_0 + \eta A^{(1)}) \\ M(A_0 + \eta A^{(2)}) \\ M(A_0 + \eta A^{(2)}) \\ \vdots \\ M(A_0 + \eta A^{(p)}) \\ M(A_0 + \eta A^{(p)}) \end{pmatrix} \qquad \tilde{s} = Q \begin{pmatrix} \lambda s \\ s - \frac{z}{\gamma} \\ \lambda s \\ s - \frac{z}{\gamma} \\ \vdots \\ \lambda s \\ s - \frac{z}{\gamma} \end{pmatrix} \qquad (4.3)
$$

The rows of the matrix Q are the generators of the cone $\mathcal{R}(V)^{\perp} \cap P$ and V is a matrix made by $2p$ row blocks equal to MB, i.e.:

$$
V = \begin{pmatrix} MB \\ MB \\ \vdots \\ MB \\ MB \end{pmatrix}
$$

and finally

$$
z_j = \max_{\delta \in \Delta} M_j D\delta
$$

where, as usual, M_j denotes the jth row vector of the matrix M.

Notice that only the coefficient matrix describing this polyhedron depends on parameter η.

Therefore we must solve the problem

$$
\eta^* = \max \eta : \max_{x \in \Sigma^*} (\tilde{M}(\eta))_i x \leq \tilde{s}_i, \quad i = 1 \ldots r \qquad (4.4)
$$

where r is the row dimension of $\tilde{M}(\eta)$.

In each of the above linear programming problems the functional to be maximized depends on the parameter η.

In [8], we have stated the following result wich allows us to solve this sort of problems.

Given the set $X = \{x : Gx \leq v\}$, $G \in R^{g \times n}$, let us denote by $S(G, v)$ the matrix in the space $R^{\nu \times (2n+m+1)}$ whose row vectors are the generators of the pointed polyhedral cone:

$$\mathcal{R} \begin{pmatrix} G' \\ -G' \\ -I \\ v' \end{pmatrix}^{\perp} \cap P$$

and let us partition the matrix $S(G, v)$ as

$$S(G, v) = [S(G, v)^{(1)} | S(G, v)(2) | S(G, v)^{(3)} | S(G, v)^{(4)}]$$

where $S(G, v)^{(1)} \in R^{\nu \times n}$, $S(G, v)^{(2)} \in R^{\nu \times n}$, $S(G, v)^{(3)} \in R^{\nu \times m}$ and $S(G, v)^{(4)} \in R^{\nu}$.

Lemma 4.1. *Let $c(\pi) \in R^n$ be a vector dependent on some parameter vector $\pi \in \Pi$ and let X the set $\{x : Gx \leq v\}$. We have:*

$$\max_{x \in X}(c(\pi), x) \leq M, \ \forall \pi \in \Pi$$

if and only if

$$(S(G, v)^{(2)} - S(G, v)^{(1)})c(\pi) \leq S(G, v)^{(4)} M, \ \forall \pi \in \Pi$$

Applying this result to the problem (4.4) we have

$$\eta^* = \max \eta : (S(M, s)^{(2)} - S(M, s)^{(1)})(\tilde{M}(\eta)_i)' \leq S(M, s)^{(4)} \tilde{s}_i \ \ i = 1 \ldots r \tag{4.5}$$

Once we have computed the set Σ^* and η^* the control laws which solve our problem are characterized as follows:

$$\begin{pmatrix} MB \\ MB \\ MB \\ MB \\ \vdots \\ MB \\ MB \\ F \end{pmatrix} u(t) \leq \begin{pmatrix} \lambda\mu(x(t))s - M(A_0 + \eta^* A^{(1)}) \\ s - \frac{z}{\gamma}M(A_0 + \eta A^{(1)}) \\ \lambda\mu(x(t))s - M(A_0 + \eta^* A^{(2)}) \\ s - \frac{z}{\gamma} - M(A_0 + \eta A^{(2)}) \\ \vdots \\ \lambda\mu(x(t))s - M(A_0 + \eta^* A^{(p)}) \\ s - \frac{z}{\gamma} - M(A_0 + \eta^* A^{(p)}) \\ \rho w \end{pmatrix} \tag{4.6}$$

where $\mu(x(t)) = \inf \alpha : Mx(t) \leq \alpha s$.

It is obviously well possible that a linear state feedback control law solve a problem P2 or P3. In this last case we are able to take account of a more general uncertain structure in the matrix A. In fact we can consider the case in which the system matrix depends on some parameter, belonging to a given set, with a given arbitrary law, i.e.

$$x(t+1) = A(\pi)x(t) + Bu(t) + D\delta(t), \ \pi \in \Pi$$

In view of lemma 4.1, lemma 3.2 can be reformulated in the following lemma 4.2.

Let us consider two nonempty convex polyhedra P_1 and P_2 described respectively by the inequalities $Gx \leq v$ and $Fx \leq w$.

Lemma 4.2. *The system $Gx \leq v$, $G \in R^{m \times n}$, is satisfied by any point of P_2 if and only if*

$$(S(F,w)^{(2)} - S(F,w)^{(1)})(G_i)' \leq S(F,w)^{(4)}v_i, \ \ i = 1 \ldots m \qquad (4.7)$$

We remark that the extended Farkas' Lemma (see [11] and [12]) furnishes an equivalent condition to those given in lemma 3.2 and in lemma 4.2, but in a form which is not handable in our context. In fact if the matrix G depends on some unknown parameter vector π, which is allowed to take values in a known set Π, and we require that the system $Gx \leq v$ is satisfied by any point of P_2, for all $\pi \in \Pi$, the condition (4.7) becomes

$$\begin{aligned} \max_{\pi \in \Pi} (S(F,w)^{(2)} - S(F,w)^{(1)})_j (G_i(\pi))' \leq S(F,w)_j^{(4)} v_i, \\ i = 1 \ldots m, \ j = 1 \ldots \nu \end{aligned} \qquad (4.8)$$

where ν is the row dimension of the matrix $S(F,w)$

This last condition is obviously useful if we want to characterize the invariance of some set with respect to a dynamical systems, with uncertain matrix A. Notice moreover that there is no assumption on the function $G(.)$ and therefore no assumption is required on $A(.)$

At this point we can state the result:

Lemma 4.3. *Given the set $U = \{u : Fu \leq w\}$, $F \in R^{g \times m}$, a control law $u(t) = Kx(t)$ such that the constraint $u(t) \in \rho U$ is satisfied for all $x(t) \in \Sigma^*$ exists if and only if the following inequalities are satisfied:*

$$[S(M,s)^{(2)} - S(M,s)^{(1)}](F_i K)' \leq \rho S(M,s)^{(4)} w_i \ \ i = 1 \ldots g \qquad (4.9)$$

Proof: apply lemma 4.2 to check the inclusion $K\Sigma^* \subseteq U$. $\square$

Let us define the set $\mathcal{K}$

$$\begin{aligned} \mathcal{K} \ = \ & \{K : (A(\pi) + BK)\Sigma^* + D\frac{\Delta}{\gamma} \subseteq \Sigma^*, \ (A(\pi) + BK)\Sigma^* \subseteq \Sigma^*, \\ & Kx \in U \ \forall x \in \Sigma^*, \ \forall \pi \in \Pi\} \end{aligned}$$

Now we can state the following:

Theorem 4.2. *Given the set Π and the set $\Sigma^* \in X_{\lambda\gamma\rho}$, the set $\mathcal{K}$ is a convex polyhedron, and a matrix K belong to $\mathcal{K}$ if and only if:*

$$\left[S(M,s)^{(2)} - S(M,s)^{(1)}\right]_j K'(F_i)' \leq \lambda \left[S(M,s)^{(4)}\right]_i v_j - z_{ij}$$

$$\left[S(M,s)^{(2)} - S(M,s)^{(1)}\right]_j K'(F_i)' \leq \left[S(M,s)^{(4)}\right]_i v_j - \frac{m_{ij}}{\gamma} - z_{ij} \quad (4.10)$$

$$\left[S(M,s)^{(2)} - S(M,s)^{(1)}\right] (F_i K)' \leq \rho S(M,s)^{(4)} w_i$$

$$i = 1 \ldots m, \quad j = 1 \ldots \nu$$

where

$$z_{ij} = \max_{\pi \in \Pi}[S(M,s)^{(2)} - S(M,s)^{(1)}]_j [A(\pi)]'(F_i)'$$

$$m_{ij} = \max_{\delta \in \Delta}[S(M,s)^{(4)}]_j F_i D \delta$$

Proof: It suffices to write the necessary and sufficient conditions such that the following inclusions hold:

$$(A(\pi) + BK)\Sigma^* \subseteq \lambda \Sigma^* \quad \forall \pi \in \Pi$$

$$(A(\pi) + BK)\Sigma^* + D\frac{\Delta}{\gamma} \subseteq \Sigma^* \quad \forall \pi \in \Pi$$

$$K\Sigma^* \subseteq U\square$$

If the inequalities (4.10) have no solution with the given set Π, we might relax our requirements, i.e. searching for a matrix K such that in the case of nominal system ($\pi = 0$) the contractivity factor λ is assured and in the case of $\pi \neq 0$ the contractivity factor is λ', $\lambda \leq \lambda' < 1$. In this case we have to solve the linear programming problem:

$$\min \lambda' \; s.\, t.$$
$$\left[S(M,s)^{(2)} - S(M,s)^{(1)}\right]_j K'(F_i)' \leq \lambda \left[S(M,s)^{(4)}\right]_i v_j$$
$$\left[S(M,s)^{(2)} - S(M,s)^{(1)}\right]_j K'(F_i)' \leq \lambda' \left[S(M,s)^{(4)}\right]_i v_j - z_{ij}$$
$$\left[S(M,s)^{(2)} - S(M,s)^{(1)}\right]_j K'(F_i)' \leq \left[S(M,s)^{(4)}\right]_i v_j - \frac{m_{ij}}{\gamma} - z_{ij} \quad (4.11)$$
$$\left[S(M,s)^{(2)} - S(M,s)^{(1)}\right] (F_i K)' \leq S(M,s)^{(4)} w_i$$
$$\lambda \leq \lambda' < 1$$

Notice that the above description of the set $\mathcal{K}$ in terms of linear inequalities gives us the possibility of choosing K such that a given performance index is maximized and/or we can request that the matrix K has some particular structure: i.e. we can solve problems in which matrix K must have a given block structure (as in decentralized control) or problems in which only the output vector is available (in this last case it suffices to express the inequalities (4.10) in terms of a matrix $K = \bar{K}C$, where C is a given matrix such that $y(t) = Cx(t)$ and $\bar{K}$ is the new unknown matrix.

5. Conclusion

We have shown that, given a system, it is possible to compute control laws which assure a degree of performance, arbitrarily close to optimal performance. The performance is evaluated in terms of three parameters, which may be associated to the rate of exponential convergence toward the origin, to the attenuation of an unknown but bounded input disturbance and to the size of the input constraining set. Moreover a technique to evaluate the degree of robustness of a given solution, with respect to parametric uncertainties of the system is described. The special case of linear state feedback control law is examined.

The approach conceptually solves problems in a very general framework. From an implementation point of view, the main drawback of the method is the required numerical effort: some experimentations are in progress to estimate its practical complexity.

References

1. F. Blanchini, "Ultimate Boundedness Control for Uncertain Discrete Time Systems via Set-Induced Lyapunov Functions", *IEEE Trans. on Automatic Control*, vol. 39, no. 2, february 1994.
2. F. M. Callier, C.A. Desoer, "Linear system theory", Springer-Verlag 1991.
3. M. A. Dahleh, I.J. Diaz-Bobillo, "Control of Uncertain Systems, A Linear Programming Approach", Prentice Hall, 1995.
4. P. d'Alessandro, M. Dalla Mora, E. De Santis, "Techniques of Linear Programming Based on the Theory of Convex Cones", *Optimization*, 20, pp. 761-777, 1989.
5. P. d'Alessandro, "The conical approach to linear programming", Gordon & Breach Publisher, 1997 (to appear).
6. P. d'Alessandro, E. De Santis, "Positiveness of Dynamic Systems with Nonpositive Coefficient Matrices", *IEEE Trans. on Automatic Control*, vol. 39, no. 1, january 1994.
7. P. d'Alessandro, E. De Santis, "General closed loop optimal solutions for linear dynamic systems with linear constraints and functional", *J. of Math. Syst., Estim. and Contr.*, vol. 6, no. 2, 1996.
8. E. De Santis "On invariant sets for constrained discrete time linear systems with disturbances and parametric uncertainties", *Research Report*, Dep. of Electrical Eng., University of L'Aquila, 1996.
9. I.J. Diaz-Bobillo, M.A. Dahleh, "State feedback l^1- optimal controllers can be dynamic", *Sys. Contr. Lett.*, vol. 19, no. 2, 1992.
10. I.J. Fialho and T.T. Georgiou, "l_1 State-Feedback Control With a Prescribed Rate of Exponential Convergence", preprint 1995.
11. J.C. Hennet "Une extension du Lemme de Farkas et son application au probleme de regulation lineaire sous constraints", *Comptes-Rendus de l'Academie des Sciences*, t. 308, serie I, pp. 415-419, 1989.
12. J.C. Hennet, "Discrete time constrained linear systems", in CONTROL AND DYNAMIC SYSTEMS, VOL. 71, pp. 157-213, Academic Press, Inc.

13. Z. Lin, Ali Saberi, "Semi-global exponential stabilization of linear discrete time systems subject to input saturation via linear feedbacks", Systems and Control Letters, 24(1995) 125-132.
14. J. S. Shamma, "Optimization of the l^∞- Induced Norm Under Full State Feedback", *IEEE Trans. on A.C.*, vol. 41, no. 4, april 1996.

Chapter 2. $\mathcal{L}^2$-Disturbance Attenuation for Linear Systems with Bounded Controls: An ARE-Based Approach

R. Suárez[1], J. Alvarez-Ramírez[1], M. Sznaier[2], C. Ibarra-Valdez[1]

[1] División de Ciencias Básicas e Ingeniería, Universidad Autónoma Metropolitana-Iztapalapa, Apdo. Postal 55-534, 09340 México D.F, México, Tel. 724-46-58, Fax 7-24-46-53. email:rsua@xanum.uam.mx

[2] Department of Electrical Engineering, The Pennsylvania State University. University Park, PA 16802. email:msznaier@frodo.ee.psu.edu

1. Introduction

Feedback stabilization of systems subject to bounded inputs has been a long-standing problem in control theory. In the case where the plant itself is linear, most approaches fall within one of the two following categories: i) Saturating a linear controller designed ignoring the control bounds and analyzing the properties of the resulting closed–loop system to establish whether or not it satisfies the performance requirements; and ii) Designing new types of nonlinear controllers, trying to account for the saturation constraint in the design process.

For linear controls with saturation, sufficient conditions for global or semiglobal stabilization, and the qualitative characterization of the region of attraction of the origin, has been studied from different viewpoints (see for instance [1, 3, 7, 14, 15, 25]). While this technique provides a simple, intuitively attractive way of handling saturation constraints, it has been shown that it might fail to find a stabilizing controller, even when one does exist. For instance, the nth-order integrator ($n \geq 3$) is an example of a linear system which can be globally stabilized using a continuous bounded function [24], but can not be globally stabilized by means of a saturated linear feedback [7]. Various approaches to the problem of feedback stabilization of linear systems by means of nonlinear controllers, different from a linear saturated controller, have been considered. Sussmann et al. [28] generalized a result due to Teel [32], prove that controllable linear systems with no open-loop eigenvalues with positive real part (**NPRPE** systems) can be globally stabilized by means of linear combinations and nested compositions of saturations of linear feedbacks. Lyapunov rather than finite-time stability can be obtained using two design techniques originally intended to obtain continuous approximations to time optimal control laws [26]. Komarov [12] proposed an approach based upon an inner estimation of the attainability sets by means of ellipsoids, leading to a feedback control law obtained by solving a matrix differential equation. Although, Komarov's control is a global stabilizer for the nth-order integrator (see [26]), for the general case the region where the control is defined is not known . Along the same lines, starting from Lyapunov–type functions ("controllability functions" [13]), Gavrilyako et al. [8] have obtained a nonlinear controller that globally stabilizes linear **NPRPE** systems, but finding this control requires solving an integral equation.

While the approaches mentioned above indicate how to design a stabilizing controller in the presence of hard actuators constraints, they do not attempt to optimize the performance of the resulting closed–loop system. On the other hand, a large number of control problems require designing a controller capable of stabilizing a given system while minimizing the worst case response to some exogenous disturbances. This problem is relevant for instance for disturbance rejection, tracking and robustness to model uncertainty (see [34] and references therein). When the exogenous disturbances are modeled as bounded energy signals and performance is measured in terms of the energy of the output, it leads to the well known $\mathcal{H}_\infty$ theory. Since its introduction, the original formulation of Zames [35] has been substantially simplified, resulting in efficient computational schemes for finding solutions. The $\mathcal{H}_\infty$ framework, combined with μ–analysis [5] has been successfully applied to a number of hard practical control problems (see for instance [22]). However, it does not allow for incorporating hard bound on the control action into the problem formulation. Recently, methodologies incorporating some classes of time domain constraints into the $\mathcal{H}\infty$ formalism have been developed [23, 20, 29, 30, 31]. However, since these techniques result in *linear controllers*, the control action constraints are satisfied only if the initial condition is confined to a neighborhood of the origin.

Recently, some results have started to emerge that consider both stability and performance. In [27], the problem of global stabilization with eigenvalue placement, by means of a continuous bounded control, was addressed using either an algebraic or a differential Riccati equation. Semiglobal stability and disturbance rejection has been addressed by Lin et al. ([16] and references therein). Additional work addressing the input-output properties of saturated systems includes [9, 4, 19].

In this paper, following an approach in the spirit of [27], we propose a procedure for synthesizing state–feedback controllers for linear systems subject to both saturation–type constraints and performance requirements expressed in the form of an $\mathcal{L}^2$ to $\mathcal{L}^2$ disturbance attenuation. Here, we obtain sufficient conditions on the bounds of the control and the disturbance in order to get the same performance as in the unconstrained case.

On the other hand, to solve the global stabilization problem, we use either an algebraic or a differential Riccati equation. The proposed control law is a continuous linear-like feedback law with state-dependent gains that globally stabilizes the plant. Moreover, this control law coincides in an ellipsoidal neighborhood of the origin with the unconstrained static state–feedback $\mathcal{H}_\infty$ controller.

2. Disturbance attenuation with bounded control problem

Consider the following linear system

$$
\begin{aligned}
\dot{x} &= Ax + Bu + Ew \\
z &= Cx + Du
\end{aligned}
\tag{2.1}
$$

where $x \in \mathbb{R}^n$ represents the state, $u \in \mathbb{R}^m$ represents the control and $w \in \mathbb{R}^r$ and $z \in \mathbb{R}^p$ represent an exogenous disturbance and the controlled output respectively. The problem that we address in this paper is the following:

Disturbance attenuation with bounded control problem (DABCP): Given the system (2.1) and real positive numbers γ, u_0, w_0; find a continuous control function $u(x)$ such that

i) $u(x)$ is bounded by u_0 ($|u_i(x)| \leq u_0, i = 1, 2, ..., m$),
ii) The system $\dot{x} = Ax + Bu(x)$ is globally asymptotically stable, and
iii) For any disturbance w, $\mathcal{L}^2$-bounded by w_0, the closed-loop system

$$
\begin{aligned}
\dot{x} &= Ax + Bu(x) + Ew \\
z &= Cx + Du(x)
\end{aligned}
\tag{2.2}
$$

has $\mathcal{L}^2$-to-$\mathcal{L}^2$ induced gain lower than γ. In other words, for any solution of (2.2), with initial condition $x(0) = 0$, the following holds:

$$
\int_0^\infty \|z(t)\|^2 dt \leq \gamma_0^2 \int_0^\infty \|w(t)\|^2 dt,
\tag{2.3}
$$

for certain $0 < \gamma_0 < \gamma$.

In the sequel, in order to solve the **DABCP** problem we will make the following assumptions:

Assumption 2.1. $\mathbf{C}^T D = 0_{p \times p}$.

Assumption 2.2. $D^T D = I$.

Assumption 2.3. The pairs (A, B) and (A, C) are stabilizable and detectable respectively.

Assumption 2.1 establishes that there are no cross terms in the formula of $\|z\|^2$, and it is required by our approach. Assumption 2.2 normalizes (without loss of generality) the penalty on the control action. Finally, Assumption 2.3 is required for the standard $\mathcal{L}^2$ to $\mathcal{L}^2$ disturbance attenuation problem (i.e. $\mathcal{H}_\infty$ problem) to have a solution.

It is well known (see [6, 17] for details) that if these assumptions are satisfied and if the following Riccati equation

$$A^T P + PA + P(\frac{1}{\gamma^2} EE^T - BB^T)P + C^T C = 0_{n \times n} \qquad (2.4)$$

admits a positive definite solution $P(\gamma) > 0$, then the linear control law $u_L(x) = -B^T P(\gamma)x$ solves the disturbance attenuation problem. Moreover, there exists γ_{opt} such that, for each $\gamma > \gamma_{opt}$, (2.4) has a unique solution $P(\gamma) > 0$ for which $A_C = A - (\frac{1}{\gamma^2} EE^T - BB^T)P(\gamma)$ is a stable matrix (this solution is called stabilizing solution).

In the remainder of this section we will obtain conditions on γ, w_0 and u_0, such that condition iii) is satisfied in $\mathcal{BE} = \{x \in \mathbb{R}^n : x^T Px \leq \beta\}$, where $\beta = \min\{u_0^2/(b_i^T Pb_i) : i = 1, ..., m\}$ is the maximum positive number which guarantees boundedness of u_L by u_0 in $\mathcal{BE}$, and b_i are the column vectors of the matrix B.

Lemma 2.1. *The feedback control function $u_L(x)$ is bounded by u_0 in $\mathcal{BE}$.*

Proof: The surface $\partial\mathcal{BE}$ is defined by the equality $x^T Px - \beta = 0$. To find the largest and smallest values of $u_{L,i}(x)$ on $\partial\mathcal{BE}$, we shall use Lagrange multipliers. The necessary condition for an extremum becomes:

$$2\bar{x}^T P + \lambda b_i^T P = 0,$$

which, by solving for $\bar{x}$, gives $\bar{x} = -\frac{1}{2} b_i \lambda$. Making use of the fact that the worst case corresponds to $-b_i^T Px \pm u_0 = 0$, we obtain $\lambda = \pm 2\frac{u_0}{b_i^T Pb_i}$, so that $\bar{x} = \pm\frac{u_0 b_i}{b_i^T Pb_i}$. Hence, at the extremum we have that:

$$\bar{x}^T P\bar{x} = \frac{u_0^2}{b_i^T Pb_i}.$$

It follows that the control input $u_{L,i}(x)$ is bounded by u_0 in the set

$$\left\{x \in \mathbb{R}^n : x^T Px \leq u_0^2/(b_i^T Pb_i)\right\}.$$

Thus, in $\mathcal{BE}$ all the control inputs $u_{L,i}(x), i = 1, ..., m$, are bounded by u_0. $\square$

Lemma 2.2. *Let $\gamma_m > \gamma_{opt}$. Then, there exists a constant $L > 0$ independent on γ, such that for all $\gamma > \gamma_m$ and for any disturbance function $w(t)$, $\mathcal{L}^2$-bounded by w_0, i.e.*

$$\|w(t)\|_2^2 \doteq \int_0^\infty \|w(t)\|^2 dt \le w_0^2 \tag{2.5}$$

the corresponding solution $x(t)$ of (2.1), with $u = u_L$ and initial condition $x(0) = 0$, is contained in the ellipsoid $\mathcal{R}_w = \{x \in \mathbb{R}^n : x^T P x \le L w_0^2\}$.

Proof: Consider the function $V(x) = x^T P x$. It can be shown ([6]) that the following equality holds:

$$\begin{aligned}
\frac{dV(x(t))}{dt} + \|z(t)\|^2 &= \gamma^2 \left[\|w(t)\|^2 - \|w(t) - w_*(t)\|\right] = \\
&= \gamma^2 \left[2\langle w(t), w_*(t)\rangle - \|w_*(t)\|^2\right]
\end{aligned}$$

Eliminating $\|z(t)\|^2$ and $\|w_*(t)\|^2$ and making $w_*(x(t)) = \frac{1}{\gamma^2} E^T P(\gamma) x(t)$, we arrive to the inequality $\frac{d}{dt} V(x(t)) \le 2\|E\|\|P(\gamma)\|\|w(t)\|\|x(t)\|$.

Now, by integration and Hölder's inequality, we get for any $t > 0$

$$x(t)^T P(\gamma) x(t) \le 2\|E\|\|P(\gamma)\|\|w\|_2\|x\|_2 \tag{2.6}$$

We shall obtain an estimate for the norm of the state when $x(0) = 0$. Denote by $A_{CL}(\gamma)$ the closed loop matrix $A - BB^T P(\gamma)$. Since $x(t) = \int_0^t e^{A_{CL}(\gamma)(t-s)} E w(s) ds$, it follows from Hausdorff-Young inequality (see [21]) that

$$\|x\|_2^2 \le \|E\|^2 \left(\int_0^\infty \|e^{A_{CL}(\gamma)s}\| ds\right)^2 \|w\|_2^2.$$

We are going to see that

$$\Theta = \sup_{\gamma \in [\gamma_m, \infty)} \int_0^\infty \|e^{A_{CL}(\gamma)s}\| ds < \infty.$$

Note that the mapping $\gamma \to P(\gamma)$ from $[\gamma_m, \infty) \cup \{\infty\}$ to the set of positive definite matrices, defines a continuous curve from $P(\gamma_m)$ to $P(\infty)$, where $P(\infty)$ is the solution to the H_2-Riccati equation $A^T P + PA - PBB^T P + C^T C = 0_{n \times n}$. Therefore, there exists $\sigma > 0$ such that $\Re e(\lambda) < -\sigma$ for all $\lambda \in \{Spec(A_{CL}(\gamma)) : \gamma \in [\gamma_m, \infty)\}$. Moreover, using the semisimple plus nilpotent decomposition of the matrices $A_{CL}(\gamma)$ (see [10], chapter 6) it can be seen that there is a constant $\eta > 0$, independent of γ, such that $\|e^{A_{CL}(\gamma)s}\| \le \eta e^{-\sigma s}$. From this it follows that $\Theta < \infty$.

Now, we claim that $\|P(\gamma)\|$ is a decreasing function of γ in the sense that $\|P(\gamma)\| \le \text{const.} \|P(\gamma_m)\|$ for all $\gamma \ge \gamma_m$. This can be seen by taking derivatives with respect to γ in the Riccati equation (2.4), and obtain

$$A_C^T P_\gamma + P_\gamma A_C = \frac{2}{\gamma^3} PEE^T P,$$

where $P_\gamma = \frac{d}{d\gamma}P$ and $A_C = A - (\frac{1}{\gamma^2}EE^T - BB^T)P$, which is an stable matrix. From the above Lyapunov equation we see that P_γ is negative definite. The claim can be readily seen from here, by considering the derivative of $g(\gamma) = \|P(\gamma)\|_F^2 = \mathrm{tr}(P(\gamma)P(\gamma))$.

Coming back to (2.6) we have for all $\gamma \in [\gamma_m, \infty)$,

$$x(t)^T P(\gamma) x(t) \leq 2\Theta \|E\|^2 \|P(\gamma_m)\| \|w\|_2^2 = L \|w\|_2^2 \leq L w_0^2. \qquad \square$$

Proposition 2.1. *Assume that u_0, w_o and γ satisfy the inequality:*

$$u_0 \geq [\max\{b_i^T P(\gamma) b_i : i = 1, ..., m\}]^{1/2} L^{1/2} w_0 \qquad (2.7)$$

where P is the solution to (2.4) and $L = 2\Theta \|E\|^2 \|P(\gamma_m)\|$. Then, condition iii) is satisfied.

Proof: The proof follows immediately from Lemmas 2.1 and 2.2, by observing that condition (2.7) implies that $\mathcal{R}_w \subset \mathcal{BE}$. $\qquad\qquad \square$

Since $b_i^T P(\gamma) b_i \ i = 1, \ldots, m$, is a monotonically decreasing function of γ, for $\gamma > \gamma_{\mathrm{opt}}$, inequality (7) implies the intuitive idea that the more demanding the performance requirement is, the stronger the control actions should be made.

Remark 2.1. The input bound u_0 estimated by inequality (7), can be possibly improved. Better estimates of u_0 could be obtained in case that tighter estimates of the trajectory peaking under the $\mathcal{L}^2$-signal $w(t)$ can be provided. $\qquad \square$

3. Global stabilization problem

In this section we extend u_L as a bounded continuous function defined in all $\mathbb{R}^n$, in order to define a feedback control which globally stabilizes (2.1) while keeps condition iii).

For $\tau \geq 1$, let $P(\tau) > 0$ be the stabilizing solution to the equation

$$A^T P + PA + P(\frac{1}{\gamma^2}EE^T - BB^T)P + \frac{1}{\tau}C^T C = 0_{n x n} \qquad (3.1)$$

Define the function $\tau(x)$ as the positive solution to the equation:

$$x^T P(\tau) x - c(\tau) = 0 \qquad (3.2)$$

where $c(\tau) = \min\{u_0^2/(b_i^T P(\tau) b_i) : i = 1, ..., m\}$. Note that $c(1) = \beta$. The feedback that we propose is a linear-like controller with state-dependent gains (variable-gains) of the form

$$u(x) = -B^T P(\tau(x))x \qquad (3.3)$$

Observe that $P(\tau(x))$ is a matrix function which is constant on the boundary of each one of the nested ellipsoidal neighborhoods of the origin (Lemmas 3.1 and 3.2) defined by

$$\mathcal{E}(\tau) = \{x \in \mathbb{R}^n : x^T P(\tau)x \le c(\tau)\}; 1 \le \tau \le \infty \qquad (3.4)$$

where, as in Lemma 2.1, the function $c(\tau)$ was chosen to guarantee boundedness of (3.3) by u_0. This procedure leads to an implicitly defined nonlinear feedback. Due to the fact that $u(x)$ is constant on the ellipsoids, we say that (3.3) is a variable-gain controller. Since $P(1) = P$, the control function $u(x)$ coincides with u_L in the ellipsoid $\partial \mathcal{E}(1) = \{x \in \mathbb{R}^n : x^T P(1)x = c(1)\}$. Next, we modify the control function $u(x)$ so that it coincides with u_L in the ellipsoidal set $\mathcal{BE} = \mathcal{E}(1)$.

Theorem 3.1. *Assume that (2.4) has a stabilizing positive solution for $\gamma_m > 0$, and inequality (2.7) holds. If (2.1)has non-positive real part open-loop eigenvalues, then the control $u(x)$ given by:*

$$u(x) = \begin{cases} -B^T P(\tau(x))x & \text{if } x \in \mathcal{E} \backslash \mathcal{BE} \\ -B^T P(1)x & \text{if } x \in \mathcal{BE} \end{cases} \qquad (3.5)$$

*where $\mathcal{E} = \cup \mathcal{E}(\tau) = \mathbb{R}^n$, solves the **DABC** problem with control constraint u_0.*

In order to prove the theorem, some definitions and preliminary results are required. Consider the positive definite matrix

$$\mathcal{R}(\tau) = P(\tau)/c(\tau),$$

for all $0 < \tau < \infty$ and redefine the τ-parameterized family of ellipsoidal sets (3.4) by:

$$\mathcal{E}(\tau) = \{x \in \mathbb{R}^n : x^T \mathcal{R}(\tau)x \le 1\} \qquad (3.6)$$

Lemma 3.1. $\mathcal{R}_\tau = \frac{d}{d\tau}\mathcal{R}(\tau)$ *is a negative definite matrix if and only if $\mathcal{E}(\tau_1) \subset Int\mathcal{E}(\tau_2)$, whenever $\tau_1 < \tau_2$.*

Proof: See [26] □

Lemma 3.2. $\mathcal{R}_\tau$ *is a negative definite matrix.*

Proof: We first prove that $P_\tau = \frac{d}{d\tau}P(\tau)$ is a negative definite matrix. Differentiating both sides of (3.1), we get

$$\frac{1}{\tau^2 \gamma^2}PEE^T P + \frac{1}{\tau^2}CC^T = A^T P_\tau + P_\tau A + P_\tau(\frac{1}{\gamma^2}EE^T - BB^T)P +$$

$$+ P(\frac{1}{\gamma^2}EE^T - BB^T)P_\tau$$

$$(3.7)$$

Denoting by $A_C = A + (\frac{1}{\gamma^2}EE^T - BB^T)P(\tau)$, (3.7) can be rewritten as:

$$A_C^T P_\tau + P_\tau A_C = \frac{1}{\tau^2\gamma^2}PEE^T P + \frac{1}{\tau^2}CC^T.$$

We have (from Theorem 3 and Lemma 6 in [6]) that A_C is a stable matrix. It follows from the detectability of the pair (A, C) that P_τ is a negative definite matrix.

For $\mathcal{R}_\tau = \frac{d}{d\tau}\mathcal{R}(\tau) = \frac{d}{d\tau}(P(\tau)/c(\tau))$, we have

$$x^T \mathcal{R}_\tau x = \frac{x^T P_\tau x c(\tau) - x^T P(\tau)x\frac{d}{d\tau}c(\tau)}{c(\tau)^2} \tag{3.8}$$

where $\frac{d}{d\tau}c(\tau)$ means the right-side derivative of $c(\tau)$. Finally, since

$$u_0^2/(b_i^T P(\tau)b_i, \quad i = 1, ..., m$$

are monotonous increasing functions of τ,

$$\frac{d}{d\tau}\frac{u_0^2}{b_i^T P(\tau)b_i} = -\frac{u_0^2 b_i^T P_\tau b_i}{(b_i^T P(\tau)b_i)^2} > 0,$$

$c(\tau)$ is also monotonous increasing. In particular, the right-side derivative of $c(\tau)$ is well defined and positive. The proposition follows from (3.8), since P_τ is a negative definite matrix.□

Combining the results of Lemmas 3.1 and 3.2 the following proposition is now evident:

Proposition 3.1. *The τ-parameterized family of ellipsoidal sets $\mathcal{E}(\tau)$ is a nested family set; that means, $\mathcal{E}(\tau_1) \subset int\ \mathcal{E}(\tau_2)$, whenever $\tau_1 < \tau_2$, with maximal element*

$$\mathcal{E} = \cup\mathcal{E}(\tau) = \lim_{\tau\to\infty}\mathcal{E}(\tau).$$

Next, we show that $u(x)$ is well defined and satisfies the hypothesis of Theorem 3.1.

Lemma 3.3. *Given $x^* \in \mathcal{E}$ and $\tau^* \in (1,\infty)$ such that $x^* \in \partial\mathcal{E}(\tau^*)$, we define $\tau(x^*) = \tau$, and if $x^* \in \mathcal{BE}$ we define $\tau(x^*) = 1$. Then, τ is a continuous function from $\mathcal{E}$ to $[1,\infty)$, differentiable in $\mathcal{E}\backslash\mathcal{BE}$.*

Proof: See [27].   □

Lemma 3.4. *The feedback control function (2.7) satisfies condition i).*

Proof: The proof follows along similar lines as in the proof of Lemma 2.1. □
Proof of Theorem 3.1: First we will prove that condition (ii) holds (therefore for this part we assume that $w = 0$). Recall that $\mathcal{BE} = \{x \in \mathbb{R}^n : x^T P(1)x \leq c(1)\}$, so therefore the boundary of $\mathcal{BE}$ is a level set of the Lyapunov function $V(x) = x^T P(1)x$. Then, $\mathcal{BE}$ is an invariant set of the linear system $\dot{x} = (A - BB^T P(1))x$. Since $u(x)$ is equal to $-B^T Px$ in $\mathcal{BE}$, it follows that all the trajectories of the closed-loop system that arrive to $\mathcal{BE}$ are driven asymptotically to the origin.

Next we show that $P(\tau) \to 0_{n \times n}$ (when $\tau \to \infty$), and therefore $\mathcal{E}(\tau)$ converges to the whole $\mathbb{R}^n$. Consider the non-negative definite matrix $P(\infty) \doteq \lim_{\tau \to \infty} P(\tau)$. Negativeness of P_τ implies that $P(\tau)$ is bounded as a function of τ. It follows that $P(\infty)$ satisfies the Riccati equation

$$A^T P + PA - P(\frac{1}{\gamma^2}EE^T - BB^T)P = 0_{n \times n} \qquad (3.9)$$

Since A has non-positive real part eigenvalues, we have from [33, 27] that $P(\infty) = 0_{n \times n}$ and therefore $\mathcal{E} = \mathbb{R}^n$.

Next, we prove that the compact set $\mathcal{BE}$ is a global attractor for the closed-loop system $\dot{x} = Ax + Bu(x)$. In $\mathcal{E} \backslash \mathcal{BE}$, $\dot{\tau}(x)$ can be calculated from (3.2), giving

$$\dot{\tau} = -\frac{2x^T P(\tau)\dot{x}}{x^T P_\tau x - \frac{d}{d\tau}c(\tau)}.$$

Then, $\mathcal{BE}$ is an attractor if the derivative of $\tau(x)$ along the closed-loop vector field satisfies $\dot{\tau} < 0$, for all $x \in \mathcal{E} \backslash \mathcal{BE}$. Since $2x^T P(\tau)\dot{x} = A_c^T P + PA_c = -(\frac{1}{\tau}C^T C - RBB^T P) < 0$ (see [11]), P_τ is negative definite, and $c(\tau)$ is an increasing function; it follows that $\dot{\tau} < 0$. Hence, the control feedback $u(x)$ drives all the points $x \in \mathcal{E} = \mathbb{R}^n$ to $\mathcal{BE}$, and therefore, to the origin. This, together with Proposition 3.1, proves that $\mathcal{BE}$ is a global attractor for the system (2.2) with the control action given by (3.5) and without disturbances. Hence condition ii) is satisfied. Finally, the proof of Theorem 3.1 follows from Lemma 3.4 (condition i)) and Proposition 2.1 (condition iii)). □

Remark 3.1. If the open-loop matrix A has eigenvalues with positive real part, the control function (3.5) is not a global stabilizer. However, if (2.4) has a stabilizing positive solution for $\gamma_m > 0$, and inequality (2.7) holds, the Disturbance Attenuation Problem with Control Constraint u_0 is solved in $\mathcal{E} = \cup \mathcal{E}(\tau)$. As in ([27]), if A has k eigenvalues with positive real part, then $\mathcal{E}$ is equal to the Cartesian product of a $k-$dimensional (open) ellipsoidal set and $\mathbb{R}^{n-k}$.

Remark 3.2. To avoid time-consuming calculations required by using a τ-parametrized algebraic Riccati equation, instead a differential Riccati equation can be considered (see [27]). Let $P(\tau)$ be the stabilizing positive definite solution to the equation

$$\frac{d}{d\tau}P = P_\tau = A^T P + PA + P(\frac{1}{\gamma^2}EE^T - BB^T)P \tag{3.10}$$

with initial condition $P(1) = P$ (the solution to (2.4)). Excepting global stability, Theorem 3.1 can be proved along the same lines. The main difference is found in Lemma 3.2, specifically in the arguments for proving that $P_\tau < 0$ for all $\tau \in (1, \infty)$: by differentiating both sides of (3.10), we get

$$\frac{d^2}{d\tau^2}P = P_{\tau\tau} = A^T P_\tau + P_\tau A + P_\tau(\frac{1}{\gamma^2}EE^T - BB^T)P + P(\frac{1}{\gamma^2}EE^T - BB^T)P_\tau.$$

This equation can be rewritten as:

$$P_{\tau\tau} = A_c^T P_\tau + P_\tau A_c.$$

Thus, P_τ satisfies a Lyapunov equation, and Lemma 3 in [18] yields, for $\tau \geq 1$

$$P_\tau = \Phi(\tau, 1)P_\tau(1)\Phi(\tau, 1),$$

where $\Phi(\tau, 1)$ denotes the transition matrix of $A_c(\tau)$. Observe that uniqueness of the solution to (2.4) implies $P_\tau(1) = -C^T C$. Therefore, it follows that P_τ is a negative definite matrix, as we wanted to prove.

The problem in considering differential Riccati equations like (3.10) is that their solutions can have finite escape times. Therefore, more conditions are required to obtain global stability. In particular, the solution (3.10) does not have finite escape times in $(1, \infty)$ if $B^T B - \frac{1}{\gamma^2}EE^T \geq 0$ ([2])

Remark 3.3. Note that $u(x)$ defined by (3.5) never saturates. This is an important property because a design based on this approach avoids adverse control behavior.

Example. Consider the following one-dimensional system ([36])

$$\dot{x} = ax + bu + w_1$$
$$z_1 = x$$
$$z_2 = u$$

The corresponding Riccati equation (2.4) becomes

$$\alpha p^2 + 2ap + 1 = 0,$$

where we introduce the γ-dependent variable $\alpha = \left(\frac{1}{\gamma^2} - b^2\right) = \frac{1-b^2\gamma^2}{\gamma^2}$. There are, in general, two solutions:

$$p_1 = p_1(\gamma) = \frac{-a + \sqrt{a^2 - \alpha}}{\alpha}; \quad p_2 = p_2(\gamma) = \frac{-a - \sqrt{a^2 - \alpha}}{\alpha} \tag{3.11}$$

we shall consider the three possible cases, depending on whether $a = 0$, $a > 0$ or $a < 0$.

1. $a = 0$ Here we have only one positive solution, namely

$$p = p(\gamma) = \frac{\gamma}{\sqrt{b^2\gamma^2 - 1}},$$

defined for $\gamma \in (\gamma_M, \infty)$, where $\gamma_M := \frac{1}{|b|}$. Moreover $p'(\gamma) < 0$ and $p(\gamma) \to \frac{1}{|b|}$ as $\gamma \to \infty$. The ellipsoid $\mathcal{BE}$ is

$$\{x : x^2 \leq \nu\} = [-\sqrt{\nu}, \sqrt{\nu}],$$

where $\nu = \nu(\gamma) = u_0^2/b^2 P(\gamma)^2$, Riccati equation (3.1) is here $\alpha p^2 + \frac{1}{\tau} = 0$, and substitution of $P = P(\tau)$ into (3.2) gives $\tau = \tau(x) = \frac{-b^2 x^2}{\alpha u_0^2}$. The control function given by (12) becomes

$$u(x) = \begin{cases} -\mathrm{sgn}(bx)u_0 & \text{if } |x| > \nu \\ \\ \dfrac{-b\gamma x}{\sqrt{b^2\gamma^2 - 1}} & \text{if } |x| \leq \nu. \end{cases}$$

2. $a < 0$. We have two solutions: $p_1(\gamma)$ (see (3.11)) which is positive only for $\gamma \in (\gamma_m, \gamma_M)$, with $\gamma_m := \frac{1}{\sqrt{a^2+b^2}}$, discontinuous at $\gamma = \gamma_M$, and negative for $\gamma > \gamma_M$. The second one is the positive stabilizing solution $p_2(\gamma)$ defined and continuous for $\gamma > \gamma_m$. Its asymptotic behavior properties are summarized in Figure 1.

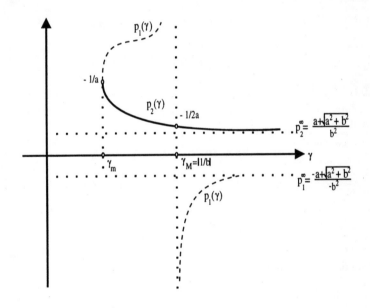

Fig. 3.1. Graph of $p(\gamma)$ for $a < 0$

3. $a > 0$. In this case $p_1(\gamma) < 0$ for every $\gamma \in (\gamma_m, \infty)$, and the second solution $p_2(\gamma)$ is negative for $\gamma_m < \gamma \leq \gamma_M$ and positive for $\gamma > \gamma_M$. Its properties are summarized in Figure 2.

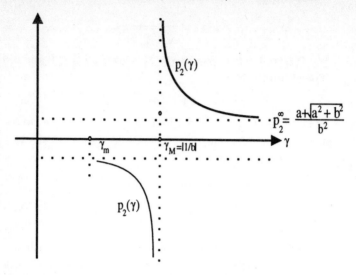

Fig. 3.2. Graph of $p(\gamma)$ for $a > 0$

For both cases $a > 0$, $a < 0$, inequality (2.7) becomes:

$$u_0 \geq \frac{2|b|}{|a|\sqrt{a^2 + b^2}}(p(\gamma))^{1/2} w_0.$$

Calculation of the bounded control $u(x)$ is algebraically more involved. It can be done as follows. Riccati equation (3.1) is here $\alpha^2 p + 2ap + 1/\tau = 0$. Substitution of $P(\tau)$ into (9) gives

$$\eta^2 + 2(\xi + 2a^2)\eta + (\xi^2 - 4a^4) = 0$$

where $\eta = \frac{\alpha}{\tau}$, $\xi = \frac{u_0^2 \alpha^2}{b^2 x^2} - 2a^2$. This quadratic equation has discriminant

$$\Delta = \frac{u_0^2 \alpha^2}{b^2 x^2} > 0.$$

From this we can obtain the positive solution $\tau = \tau_+(x)$ of (9) and calculate the control by means of (12).

4. Conclusions

A large number of control problems involve designing a controller that globally stabilizes a system, while at the same time, satisfies multiple design specifications. In this paper, following an approach in the spirit of [27], we propose a procedure for synthesizing state–feedback controllers for linear systems subject to both saturation–type constraints and performance requirements expressed in the form of an $\mathcal{L}^2$ to $\mathcal{L}^2$ disturbance attenuation. The proposed control law is a continuous linear-like feedback law with state-dependent gains that globally stabilizes the plant. These state–dependent gains are implicitly defined from the solution of an $\mathcal{H}_\infty$–type algebraic or differential Riccati equation and a nonlinear scalar equation.

References

1. J. Alvarez-Ramírez, R. Suárez and J. Alvarez, Semi-global stabilization of multi-input linear systems with saturated linear state feedback, *Syst. & Contr. Lett.*, **23** (1994) 247-254.
2. P. Bolzern, P. Colaneri and G. de Nicolao, Finite escape-times and convergence properties of the H_∞-differential Riccati equation. *13th. Triennial World Congress IFAC* San Francisco CA, USA (1996) 161-166.
3. C. Burgat, S. Tarbouriech and M. Klai, Continuous-time saturated state feedback regulators: theory and design, *Int. J. Syst, Sci.*, **25**(1994) 315-336.
4. Y. Chitour, W. Liu and E. Sontag, On the continuity and incremental-gain properties of certain saturated linear feedback loops, *Int. J. Robust Nonl. Contr.*, **5**(1995) 413-440.
5. J. Doyle, Analysis of Feedback Systems with Structured Uncertainties, *IEE Proceedings, Part D*, **129**(1982) 242–250.
6. J.C. Doyle, K. Glover, P.P. Khargonekar and B. Francis, State-space solutions to the standard $\mathcal{H}^2$ and $\mathcal{H}^\infty$ control problems, *IEEE Trans. Automat. Contr.*, **34**(1989) 831-849.
7. A.T. Fuller, In-the-large stability of relay and saturating control systems with linear controllers, *Int. J. Control*, **10**(1969) 457-480.
8. V.M. Gavrilyako, V.I. Korobov and G.M. Sklyar, Designing a bounded control of dynamic systems in entire space with the aid of a controllability function, *Automation and Remote Contr.*, **11**(1986) 1484-1490.
9. J.W. Helton and W. Zhan, Piecewise Riccati equations and the Bounded Real Lemma, *Proc. 32th. IEEE Conference on Decision and Contr.*, (1993) 196-201.
10. M. Hirsh & S. Smale, Differential equations, dynamical systems and linear algebra. Academic Press (1974).
11. A. Isidori, Control Systems (in Italian), Roma: SIDEREA (1993).
12. V.A. Komarov, Design of constrained controls for nonautonomous linear systems, *Automation and Remote Contr.*, **10**(1985) 1280-1286.
13. V.I. Korobov, A general approach to the solution of the bounded control synthesis problem in a controllability problem, *Math. USSR Sbornik*, **37**(1985) 535-539.
14. S. Lefschetz, Stability of Nonlinear Control Systems. New York: Academic Press, (1965).
15. Z. Lin and A. Saberi, Semi-global exponential stabilization of linear systems subject to "input saturation" via linear feedbacks, *Syst. & Contr. Letters*, **21**(1993) 225-239.
16. Z. Lin, A. Saberi and A.R. Teel, Simultaneous $\mathcal{L}_p$-stabilization and internal stabilization of linear systems subject to input saturation: State feedback case, *Syst. & Contr. Letters*, **25** (1995) 219-226.

17. I.R. Petersen, Disturbance attenuation and $\mathcal{H}^\infty$ optimization: A design method based on the algebraic Riccati equation, *IEEE Trans. Automat. Contr.,* **32**(1987) 427–429.

18. M.A. Poubelle, R.R. Bitmead and M.R. Gevers, Fake algebraic Riccati techniques and stability, *IEEE Trans. Autom. Contr.,* **33**, 4, (1988) 379-381.

19. B.G. Romanchuk, On the computations of the induced $\mathcal{L}_2$ norm of single input linear systems with saturation, *Proc. 33th. IEEE Conference on Decision and Contr.,* (1994) 1427-1432.

20. H. Rotstein and A. Sideris, $\mathcal{H}_\infty$ Optimization with Time–Domain Constraints, *IEEE Trans. Autom. Contr.,* **39**, 4, (1994) 762–779.

21. E.M. Stein, G. Weiss, Fourier Analysis in Euclidean Space, *Princeton Univ. Press,* Princeton NJ (1971).

22. S. Skogestad, M. Morari and J. Doyle, Robust Control of Ill–Conditioned Plants: High–Purity Distillation, *IEEE Trans. Autom. Contr.,* **33**, 12, (1988) 1092–1105.

23. A. Sideris and H. Rotstein, Single Input–Single Output $\mathcal{H}_\infty$ Control with Time Domain Constraints, *Automatica,* **29**, 4, (1993) 969–984.

24. E.D. Sontag and H.J. Sussmann, Nonlinear output feedback design for linear systems with saturating controls, *Proc. 29th. IEEE Conference on Decision and Contr.,* (1990) 3414-3416.

25. R. Suárez, J. Alvarez, and J. Alvarez, Regions of Attraction of Closed Loop Linear Systems with Saturated Linear Feedback *Journal of Math. Syst., Est., Contr.,* (To appear).

26. R. Suárez, J. Solis-Daun, and J. Alvarez, Stabilization of linear control systems by means of bounded continuous nonlinear feedback control, *Syst. & Contr. Letters,* **23**(1994) 403-410.

27. R. Suárez, J. Alvarez-Ramírez and J. Solis-Daun, Linear systems with bounded inputs: Global stabilization with eigenvalue placement, *Int. J. Robust Nonl. Contr., (to appear).*

28. H.J. Sussmann, E. Sontag and Y. Yang, A general result on stabilization of linear systems using bounded control, *IEEE Trans. Autom. Contr.,* **39**(1994) 2411-2424.

29. M. Sznaier, "Mixed $l^1/\mathcal{H}_\infty$ Controllers for SISO Discrete–Time Systems," *Syst. & Contr. Letters,* **23**(1994) 487–492.

30. M. Sznaier, "A Mixed l_∞/H_∞ Optimization Approach to Robust Controller Design," *SIAM Journal Opt. Contr.,* to appear (1995).

31. M. Sznaier and F. Blanchini, "Robust Control of Constrained Systems via Convex Optimization," *International Journal of Robust and Nonlinear Control,* to appear (1995).

32. A.R. Teel, Global stabilization and restricted tracking for multiple integrators with bounded controls, *Syst. & Contr. Letters,* **18**(1992) 165-171.

33. A.R. Teel, Semiglobal stabilizability of linear null controllable systems with input nonlinearities, *IEEE Trans. Autom. Contr.,* **40**, 1, (1995) 96–100.

34. M. Vidyasagar, Optimal Rejection of Persistent Bounded Disturbances, *IEEE Trans. Autom. Contr.,* **31**, 6, (1986) 527–535.

35. G. Zames, Feedback and Optimal Sensitivity: Model Reference Transformations, Multiplicative Seminorms and Approximate Inverses, *IEEE Trans. Autom. Contr.* **26**, 4, (1981) 301–320.

36. K. Zhou, J. Doyle and K. Glover, Robust and Optimal Control, *Prentice-Hall* 1996.

Chapter 3. Stability Analysis of Uncertain Systems with Saturation Constraints

Anthony N. Michel and Ling Hou

Department of Electrical Engineering, University of Notre Dame, Notre Dame, IN 46556, U.S.A.

1. Introduction

In this paper, we study robust stability of linear systems with parameter uncertainties and subject to state saturation. We first investigate the stability properties of systems operating on the closed unit hypercube, described by

$$\dot{x}(t) = h\Big(Ax(t)\Big) \tag{1.1}$$

where $x \in D^n = \Big\{x = (x_1, \cdots x_n)^T \in R^n : -1 \leq x_i \leq 1, i = 1, \cdots, n\Big\}$, $A = [a_{ij}] \in R^{n \times n}$, $\dot{x} = dx/dt$, $t \geq 0$,

$$h(Ax) = \Big[h_{x_1}\Big(\sum_{j=1}^{n} a_{1j}x_j\Big), \cdots, h_{x_n}\Big(\sum_{j=1}^{n} a_{nj}x_j\Big)\Big]^T \tag{1.2}$$

and

$$h_{x_i}\Big(\sum_{j=1}^{n} a_{ij}x_j\Big) = \begin{cases} 0, & |x_i| = 1 \text{ and } \Big(\sum_{j=1}^{n} a_{ij}x_j\Big)x_i \geq 0 \\ \sum_{j=1}^{n} a_{ij}x_j, & \text{otherwise.} \end{cases} \tag{1.3}$$

We assume that in system (1.1), the matrix A is known to belong to an interval matrix, i.e., $A \in [A^n, A^M]$. (An *interval matrix* $[A^n, A^M]$ with $A^m = [a_{ij}^m] \in R^{n \times n}$, $A^M = [a_{ij}^M] \in R^{n \times n}$, and $a_{ij}^m \leq a_{ij}^M$ for all i and j, is defined by $[A^n, A^M] \triangleq \{C = [c_{ij}] \in R^{n \times n} : a_{ij}^m \leq c_{ij} \leq a_{ij}^M, 1 \leq i, j \leq n\}$.)

In this paper, we also investigate the stability properties of systems with partial state saturation, described by differential equations of the form

$$\begin{cases} \dot{x} = A_{11}x + A_{12}y \\ \dot{y} = h(A_{21}x + A_{22}y) \end{cases} \tag{1.4}$$

where $n \geq m$, $x \in R^{(n-m)}$, $y \in D^m = \Big\{y = (y_1, \cdots, y_m)^T \in R^m : -1 \leq y_i \leq 1, i = 1, \cdots, m\Big\}$, A_{11}, A_{12}, A_{21}, and A_{22} are real matrices of appropriate dimension, $\dot{x} = dx/dt$, $\dot{y} = dy/dt$, $t \geq 0$,

$$h(A_{21}x + A_{22}y) = \left[h_{y_1}\left(\sum_{j=1}^{n-m} c_{1j}x_j + \sum_{k=1}^{m} d_{1k}y_k\right), \cdots, h_{y_m}\left(\sum_{j=1}^{n-m} c_{mj}x_j + \sum_{k=1}^{m} d_{mk}y_k\right)\right]^T$$

(1.5)

and

$$h_{y_i}\left(\sum_{j=1}^{n-m} c_{ij}x_j + \sum_{k=1}^{m} d_{ik}y_k\right)$$

$$= \begin{cases} 0, & |y_i| = 1 \text{ and } \left(\sum_{j=1}^{n-m} c_{ij}x_j + \sum_{k=1}^{m} d_{ik}y_k\right)y_i > 0 \\ \sum_{j=1}^{n-m} c_{ij}x_j + \sum_{k=1}^{m} d_{ik}y_k, & \text{otherwise,} \end{cases}$$

(1.6)

where $A_{21} = [c_{ij}] \in R^{m \times (n-m)}$ and $A_{22} = [d_{ij}] \in R^{m \times m}$. We assume that in system (1.4), the matrix

$$A \triangleq \begin{bmatrix} A_{11} & A_{12} \\ A_{21} & A_{22} \end{bmatrix} \in [A^m, A^M].$$

2. Existence of Solutions

Before proceding, it is necessary to consider questions of existence, uniqueness, and continuation of solutions of (1.1), resp., (1.4). In the following we consider (1.1) without loss of generality. We will first show that for any $x(t_0) \in D^n$, there is a local solution for (1.1).

(A) From well-known results (e.g., [2]), we know that for any $x(t_0)$ in the interior of D^n, there exists $\delta_{x(t_0)} > 0$ such that $e^{A(t-t_0)}x(t_0)$ is in the interior of D^n for all $t \in [t_0, t_0 + \delta_{x(t_0)})$. Therefore, the solution of equation (1.1) exists in this case and is unique and continuous on $[t_0, t_0 + \delta_{x(t_0)})$.

(B) For any $x(t_0) \in \partial D^n$, let $\sigma: \{1, \cdots, n\} \to \{1, \cdots, n\}$ be a permutation such that for some l, k and m such that $0 \le k \le l \le m \le n$, we have

$$\dot{x}_{\sigma(i)}(t_0) = \sum_{j=1}^{n} a_{\sigma(i)j}x_j(t_0), \qquad |x_{\sigma(i)}(t_0)| < 1, \quad i \in \{1, \cdots, k\};$$

$$\dot{x}_{\sigma(i)}(t_0) = \sum_{j=1}^{n} a_{\sigma(i)j}x_j(t_0), \qquad |x_{\sigma(i)}(t_0)| = 1,$$
$$\text{and } \left(\sum_{j=1}^{n} a_{\sigma(i)j}x_j(t_0)\right)x_{\sigma(i)}(t_0) < 0, \quad i \in \{k+1, \cdots, l\};$$

$$\dot{x}_{\sigma(i)}(t_0) = \sum_{j=1}^{n} a_{\sigma(i)j}x_j(t_0) = 0, \qquad |x_{\sigma(i)}(t_0)| = 1, \quad i \in \{l+1, \cdots, m\}; \text{ and}$$

$$\dot{x}_{\sigma(i)}(t_0) = 0, \qquad |x_{\sigma(i)}(t_0)| = 1, \text{ and}$$
$$\left(\sum_{j=1}^{n} a_{\sigma(i)j}x_j(t_0)\right)x_{\sigma(i)}(t_0) > 0, \quad i \in \{m+1, \cdots, n\}.$$

We define $\{i, \cdots, j\} = \emptyset$, if $j < i$.

Without loss of generality, we assume in the following that $\sigma(i) = i$. Let $x_I = \{x_1, \cdots, x_l\}$, $x_{II} = \{x_{l+1}, \cdots, x_n\}$, and $x = (x_I^T, x_{II}^T)^T$, and let $x(t_0)$, $t = t_0$ be initial date. We have

$$\dot{x}_I = A_{I,I} x_I + A_{I,II} x_{II}$$
$$\dot{x}_{II} = 0$$

where $A_{I,I}$ and $A_{I,II}$ are appropriate submatrices of A.

For $i \in \{1, \cdots, k\}$, let $d_i = 1 - |x_i(t_0)|$.

For $i \in \{k+1, \cdots, l\}$, there exists $d_i > 0$ such that $(\sum_{j=1}^n a_{\sigma(i)j} x_j) x_{\sigma(i)}(t_0) < 0$, for all x such $\|x - x(t_0)\| < d_i$. Suppose that $x(t)$ with $t \in [t_0, t_0+\tau), \tau > 0$, and $\|x(t) - x(t_0)\| < d_i$, is a solution of equation (1.1). Then, $|x_i(t)|$ is strictly decreasing and it must be true that $|x_i(t)| < 1$ for all $t \in (t_0, t_0 + \tau)$.

For $i \in \{l+1, \cdots, m\}$, consider the second-order *right derivative* of x_i with respect to t,

$$\frac{d^2}{dt^2} x_i = \frac{d}{dt} \sum_{j=1}^n a_{ij} x_j = [a_{i1}, \cdots, a_{il}](A_{I,I} x_I + A_{I,II} x_{II}).$$

If $\frac{d^2 x_i(t_0)}{dt^2} x_i(t_0) < 0$, then there exists $d_i > 0$ such that the right side of the above equation is negative for all x such that $\|x - x(t_0)\| < d_i$. If $x(t)$ with $t \in [t_0, t_0 + \tau), \tau > 0$, and $\|x(t) - x(t_0)\| < d_i$ is a solution of equation (1.1), then $\frac{dx_i(t_0)}{dt}$ is of opposite sign of $x_i(t_0)$. Hence, $|x_i(t)|$ is strictly decreasing and it must be true that $|x_i(t)| < 1$ for all $t \in (t_0, t_0 + \tau)$. Similarly, if $\frac{d^2 x_i(t_0)}{dt^2} x_i(t_0) > 0$, then there exists $d_i > 0$ such that the right side of the above equation is positive for all x such that $\|x - x(t_0)\| < d_i$. If $x(t), t \in [t_0, t_0 + \tau), \tau > 0$, and $\|x(t) - x(t_0)\| < d_i$ is a solution of equation (1.1), then it must be true that $x_i(t) = x_i(t_0)$ and $(\sum_{j=i}^n a_{ij} x_j(t)) x_i(t) > 0$ for all $t \in (t_0, t_0 + \tau)$. If $\frac{d^\lambda}{dt^\lambda} x_i(t_0) = 0$ then consider $\frac{d^{\lambda+1}}{dt^{\lambda+1}} x_i(t_0) = [a_{i1}, \cdots, a_{ik}](A_{I,I})^{\lambda-1}(A_{I,I} x_I + A_{I,II} x_{II})$, $\lambda = 2, \cdots, l$, and apply a similar argument as above. If $\frac{d^\lambda}{dt^\lambda} x_i(t_0) = [a_{i1}, \cdots, a_{ik}](A_{I,I})^{\lambda-2} A_{I,I} x_I + A_{I,II} x_{II}) = 0$ for all $\lambda = 1, \cdots, l$, then $\frac{d^\lambda x_i(t_0)}{dt^\lambda} x_i(t_0) = 0$ for all λ and we let $d_i = 1$.

For $i \in \{m+1, \cdots, n\}$, there exists $d_i > 0$ such that $(\sum_{j=1}^n a_{\sigma(i)j} x_j) x_{\sigma(i)}(t_0) > 0$ for all x such that $\|x - x(t_0)\| < d_i$. If $x(t)$ with $t \in [t_0, t_0 + \tau), \tau > 0$, and $\|x(t) - x(t_0)\| < d_i$ is a solution of equation (1.1), then it has to be true that $x_i(t) = x_i(t_0)$ and $(\sum_{j=1}^n a_{\sigma(i)j} x_j(t)) x_{\sigma(i)}(t) > 0$, for all $t \in [t_0, t_0 + \tau)$. We now determine the solution of (1.1). Without loss of generality, we assume that there exist r and s such that $l \leq r \leq s \leq m$ and such that for $i \in \{l+1, \cdots, r\}$, there exists $\lambda_i > 1$ such that $\frac{d^\lambda x_i(t_0)}{dt^\lambda} = 0$ when $\lambda < \lambda_i$ and $\frac{d^{\lambda_i} x_i(t_0)}{dt^{\lambda_i}} x_i(t_0) < 0$; for $i \in \{r+1, \cdots, s\}$, $\frac{d^\lambda x_i(t_0)}{dt^\lambda} x_i(t_0) = 0$ for all λ; and for

$i \in \{s+1, \cdots, m\}$, there exists $\lambda_i > 1$ such that $\frac{d^\lambda x_i(t_0)}{dt^\lambda} = 0$ when $\lambda < \lambda_i$ and $\frac{d^{\lambda_i} x_i(t_0)}{dt^{\lambda_i}} x_i(t_0) > 0$. Let $\mathrm{M} = max\{|\sum_{j=1}^{n} a_{ij}x_j|, i = 1, \cdots, n, x \in D^n\}$ and let $d = min\{d_i, i = 1, \cdots, n\}$. Let $\tilde{x}_I = \{x_1, \cdots, x_r\}^T$, $\tilde{x}_{II} = \{x_{r+1}, \cdots, n\}^T$, $\tilde{A}_{I,I} = [a_{ij}]_{1 \le i \le r, \, 1 \le j \le r}$, $\tilde{A}_{I,II} = [a_{ij}]_{1 \le i \le r, \, r+1 \le j \le n}$, and

$$\tilde{x}_I(t) = e^{\tilde{A}_{I,I}(t-t_0)}\tilde{x}_I(t_0) + \int_0^{t-t_0} e^{\tilde{A}_{I,I}\tau} d\tau \tilde{A}_{I,II}\tilde{x}_{II}(t_0)$$
$$\tilde{x}_{II}(t) = \tilde{x}_{II}(t_0)$$

for all $t \in [t_0, t_0 + d/M)$.

We now verify that $x(t, t_0, x(t_0)) = (\tilde{x}_I^T(t), \tilde{x}_{II}^T(t))^T$, $t \in [t_0, t_0+d/M)$ is a solution of equation (1.1). It follows from the foregoing argument that in $(t_0, t_0+d/M)$, $|x_i(t)| < 1$ for all $i \in \{1, \cdots, r\}$, and $\dot{x}_i = h_{x_i}(\sum_{j=1}^{n} a_{ij}x_j) = \sum_{j=1}^{n} a_{ij}x_j$. For $i \in \{r+1, \cdots, s\}$, we have that

$$\sum_{j=1}^{n} a_{ij}x_j(t)$$
$$= [a_{i1}, \cdots, a_{ir}]\left(e^{\tilde{A}_{I,I}(t-t_0)}\tilde{x}_I(t_0) + \int_0^{t-t_0} e^{\tilde{A}_{I,I}\tau} d\tau \tilde{A}_{I,II}\tilde{x}_{II}(t_0)\right)$$
$$\qquad + [a_{i(r+1)}, \cdots, a_{in}]\tilde{x}_{II}(t_0)$$
$$= [a_{i1}, \cdots, a_{ir}]\left(\sum_{\lambda=0}^{\infty} \frac{1}{\lambda!}(\tilde{A}_{I,I})^\lambda(t-t_0)^\lambda \tilde{x}_I(t_0)\right.$$
$$\qquad \left. + \sum_{\lambda=1}^{\infty} \frac{1}{\lambda!}(\tilde{A}_{I,I})^{\lambda-1}(t-t_0)^\lambda \tilde{A}_{I,II}\tilde{x}_{II}(t_0)\right) + [a_{i(r+1)}, \cdots, a_{in}]\tilde{x}_{II}(t_0)$$
$$= [a_{i1}, \cdots, a_{in}]x(t_0) + \sum_{\lambda=1}^{\infty} \frac{1}{\lambda!}[a_{i1}, \cdots, a_{ir}](\tilde{A}_{I,I})^{\lambda-1}$$
$$\qquad \times \left(\tilde{A}_{I,I}\tilde{x}_I(t_0) + \tilde{A}_{I,II}\tilde{x}_{II}(t_0)\right)(t-t_0)^\lambda$$
$$= \sum_{\lambda=0}^{\infty} \frac{1}{\lambda!} \frac{d^{\lambda+1} x_i(t_0)}{dt^\lambda} = 0.$$

For $i \in \{s+1, \cdots, n\}$, it follows from the choice of d that $\dot{x}_i = h_{x_i}(\sum_{j=1}^{n} a_{ij}x_j) = 0$. Therefore, $x(t, t_0, x(t_0)) = (\tilde{x}_I^T(t), \tilde{x}_{II}^T(t))^T$ is a solution of equation (1.1) in the time interval $[t_0, t_0 + d)$.

Suppose there is another solution $x'(t, t_0, x(t_0))$ of equation (1.1) with $x'(t_0) = x(t_0)$, $t \in [t_0, t_0+\tau)$, $\tau > 0$. In the argument above, we have shown that in a neighborhood of $x(t_0)$, $h_{x_i}(\sum_{j=1}^{n} a_{ij}x_j)$ has to equal $\sum_{j=1}^{n} a_{ij}x_j$ for $i \in \{1, \cdots, r\}$ and $h_{x_i}(\sum_{j=1}^{n} a_{ij}x_j)$ has to be 0 for $i \in \{s+1, \cdots, n\}$. For the same reasons, we have that $h_{x_i}(\sum_{j=1}^{n} a_{ij}x'_j) = \sum_{j=1}^{n} a_{ij}x'_j$ for $i \in \{1, \cdots, r\}$ and $h_{x_i}(\sum_{j=1}^{n} a_{ij}x'_j) = 0$ for $i \in \{s+1, \cdots, n\}$. For $i \in \{r+1, \cdots, s\}$, $h_{x_i}(\sum_{j=1}^{n} a_{ij}x_j) = \sum_{j=1}^{n} a_{ij}x_j = 0$ along $x(t, t_0, x(t_0))$. Then

$$h_{x_i}\left(\sum_{j=1}^{n} a_{ij}x_j\right) - h_{x_i}\left(\sum_{j=1}^{n} a_{ij}x'_j\right)$$
$$= \begin{cases} 0, & \text{if } h_{x_i}\left(\sum_{j=1}^{n} a_{ij}x'_j\right) = 0 \\ \sum_{j=1}^{n} a_{ij}(x_j - x'_j), & \text{if } h_{x_i}\left(\sum_{j=1}^{n} a_{ij}x'_j\right) = \sum_{j=1}^{n} a_{ij}x'_j. \end{cases}$$

Therefore,

$$\left| h_{x_i}\left(\sum_{j=1}^{n} a_{ij}x_j\right) - h_{x_i}\left(\sum_{j=1}^{n} a_{ij}x_j'\right) \right| \le K\|x(t) - x'(t)\|$$

for all $i = 1, \cdots, n$, where $K = max\{|a_{ij}| : 1 \le i, j \le n\}$ Then,

$$\|\dot{x}(t) - \dot{x}'(t)\| = \|h(Ax(t)) - h(Ax'(t))\| \le nK\|x(t) - x'(t)\|.$$

Integrating both sides of the above inequality from t_0 to $t_0 + min(\tau, d/M)$, and applying the Gronwall Inequality (see, e.g., [2], p.43), we conclude that $x'(t) = x(t)$, for all $t \in [t_0, t_0 + min(\tau, d/M))$, i.e., $x(t)$ is the unique solution of equation (1.1).

Combining (A) and (B) above, we conclude that there exists a unique *local* solution of equation (1.1) for any initial point in D^n. By continuation (see, e.g., [2]), we can extend the local solution to $\cup_0^\infty [t_i, t_{i+1}) = [t_0, \infty)$ for any initial condition. On each $[t_i, t_{i+1})$, the solution is continuously differentiable (at t_i, we mean the right derivative), so that $h(Ax(t))$ is continuous in (t_i, t_{i+1}) and continuous from the right at t_i.

3. Systems Operating on a Closed Unit Hypercube

Before presenting the main results, we introduce some necessary notation.

A square matrix is said to be *positive definite* if it is symmetric and if all its eigenvalues are positive. A matrix $P = [p_{ij}] \in R^{n \times n}$ is said to be *diagonally dominant*, if

$$p_{ii} \ge \sum_{j=1, \, j\ne i}^{n} |p_{ji}|, \quad i = 1, \cdots, n, \tag{3.1}$$

where $|\cdot|$ denotes absolute value.

Let

$$\Omega = \Big\{ (I, J, I', J') \, : \, I \cup J = I' \cup J' = \{1, 2, \cdots, n\}$$

$$\text{and } I \cap J = I \cap J' = \emptyset \Big\}. \tag{3.2}$$

For every $(I, J, I', J') \in \Omega$, we let $A_{I'J'}^{IJ} = [a_{ij}^{IJI'J'}] \in R^{n \times n}$, where

$$a_{ij}^{IJI'J'} = \begin{cases} a_{ij}^M & \text{if } (i,j) \in I' \times I \cup J' \times J \\ a_{ij}^m & \text{if } (i,j) \in I' \times J \cup J' \times I \end{cases} \tag{3.3}$$

with the convention that $I' \times I = \emptyset$ when $I' = \emptyset$ or when $I = \emptyset$.

A point $x_e \in R^n$ is called an equilibrium of system (1.1) if $h(Ax_e)=0$. In particular, $x_e=0$ is an equilibrium of system (1.1). For the various definitions of stability of an equilibrium in the sense of Lyapunov, and for stability results, refer, e.g., to [1]. In the present context, the equilibrium $x_e = 0$ of system (1.1) is said to be globally asymptotically stable if $x_e=0$ is stable and the domain of attraction of $x_e=0$ is all of D^n, since system (1.1) is defined only on D^n.

The following lemma [2] is required in establishing our results.

Lemma 3.1. *Assume that for matrix $A=A_0$ there exists $V: D^n \to R$ satisfying the following assumptions:*

(A-1) *V is positive definite on D^n, $\frac{\partial V}{\partial x}(x)=\left[\frac{\partial V}{\partial x_1}(x),\cdots,\frac{\partial V}{\partial x_n}(x)\right]^T$ exists for all $x \in D^n$, and $\left[\frac{\partial V}{\partial x}(x)\right]^T Ax$ is negative definite on D^n.*
(A-2) *For all $x \in D^n$ it is true that*

$$\left[\frac{\partial V}{\partial x}(x)\right]^T h(Ax) \leq \left[\frac{\partial V}{\partial x}(x)\right]^T Ax. \tag{3.4}$$

Then the equilibrium $x_e=0$ of system (1.1) with $A=A_0$ is globally asymptotically stable.

Proof: In view of (1.2) and (1.3), it follows readily that along the solutions of (1.1), $h\left(Ax(t)\right)$ is continuous from the right in the sense that $\lim_{t \to t_1, t > t_1} h\left(Ax(t)\right) = h\left(Ax(t_1)\right)$ for any t_1. Therefore, along the solutions of system (1.1), the right-hand derivative of $V(x(t))$, given by

$$\dot{V}(x(t))=\left[\frac{\partial V}{\partial x}(x(t))\right]^T h\left(Ax(t)\right)$$

exists for all $t \geq t_0=0$. Now by assumption (A-2), $\dot{V}(x(t)) \leq [\frac{\partial V}{\partial x}(x(t))]^T Ax(t) < 0$ for all $x \neq 0$. This shows that $x_e=0$ is the only equilibrium of system (1.1).

We next show that $x_e = 0$ is asymptotically stable. Since V is positive definite and $\dot{V}$ is negative definite on D^n, there exist continuous and strictly increasing functions $\phi_i : [0,1] \to R$ such that $\phi_i(0) = 0$ (i.e., ϕ_i belongs to *class K*), $i = 1,2,3$, and

$$\phi_1(\|x\|) \leq V(x) \leq \phi_2(\|x\|), \text{ for all } x \in D^n$$

and

$$\dot{V}(x) \leq \left[\frac{\partial V}{\partial x}(x)\right]^T Ax \leq -\phi_3(\|x\|), \text{ for all } x \in D^n,$$

where $\|\cdot\|$ denotes any one of the equivalent norms on R^n. It follows from one of the principal Lyapunov stability results (see, e.g., [1]) that the equilibrium $x_e = 0$ of system (1.1) is asymptotically stable.

To conclude the proof, we show that $x_e = 0$ is *globally* asymptotically stable. First, since $x_e = 0$ is asymptotically stable, we note that there exists $d > 0$ such that $\|x(t)\| < \frac{1}{2}$ for all $t \geq t_0 = 0$ whenever $\|x(0)\| < d$ and $\lim_{t \to \infty} x(t) = 0$. Now let x_0 denote *any initial point* in D^n for the solution $x(t) = x(t, x_0, 0)$. Then there must exist $t_1 > 0$ such that $\|x(t_1)\| < d$, and hence, $\|x(t)\| < \frac{1}{2}$ for all $t \geq t_1$ and $\lim_{t \to \infty} x(t) = 0$; for otherwise, we must have $\|x(t)\| \geq d$ for all t, and then

$$V(x(t)) = V(x(0)) + \int_0^t \dot{V}(x(\tau))d\tau$$

$$\leq V(x(0)) - \int_0^t \phi_3(\|x(\tau)\|)d\tau \leq V(x(0)) - t\phi_3(d)$$

and when t is sufficiently large, then $V(x(t)) < 0$. But this is a contradiction, since by assumption (A-1), $V(x(t)) \geq 0$.

We have shown that $x_e = 0$ of system (1.1) is asymptotically stable and the domain of attraction of $x_e = 0$ is all of D^n.   □

The following theorem follows now directly from Lemma 3.1.

Theorem 3.1.   *If there exists $V_A : D^n \to R$ satisfying assumptions (A-1) and (A-2) for every $A \in [A^m, A^M]$, then the equilibrium $x_e = 0$ of system (1.1) is* globally *asymptotically stable.*   □

We note that in order to apply Theorem 3.1, we need to check every matrix $A \in [A^m, A^M]$. Without some specific criterion, this is in general impossible to verify. In the case of second-order linear systems, however, there is a simple necessary and sufficient condition to accomplish this. This criterion will enable us to establish in Section 4 a set of necessary and sufficient conditions for global asymptotic stability for second-order systems given by (1.1). For higher order linear systems, we establish the result given below. In doing so, we require the following preliminary results.

Lemma 3.2 (2).   *For any $x \in R^n$ and any $A \in [A^m, A^M]$, Ax is in the convex set with vertices $A_{I'J'}^{IJ}x$, where $(I, J, I', J') \in \Omega$, and where Ω and $A_{I'J'}^{IJ}$ are given in (3.2) and (3.3).*

□

Lemma 3.3 (4).   *Let $P \in R^{n \times n}$ be a positive definite, diagonally dominant matrix and let $V(x) = x^T Px$ where $x \in D^n$. Then, along the solutions of system (1.1), it is true that*

$$\dot{V}(x(t)) = \left[\frac{\partial V}{\partial x}(x)\right]^T h(Ax) \le \left[\frac{\partial V}{\partial x}(x)\right]^T Ax \qquad (3.5)$$

for any $x \in D^n$ and $A \in R^{n \times n}$.

Proof: There are two types of points in D^n to consider:

(i) For any $x \in D^n$ such that $h(Ax) = Ax$, we have

$$\dot{V}(x) = \left[\frac{\partial V}{\partial x}(x)\right]^T, \ Ax = x^T(PA + A^T P)x = -x^T Qx$$

and therefore relation (3.5) is satisfied.

(ii) Assume that $h(Ax) \ne Ax$ so that $x \in \partial D^n$. Let $\sigma : \{1, \cdots, n\} \to \{1, \cdots, n\}$ be a permutation such that for some $1 \le m < n$, it is true that

$$\dot{x}_{\sigma(i)} = \sum_{j=1}^n a_{\sigma(i)j} x_j, \qquad i = 1, \cdots, m \qquad (3.6)$$

and

$$\dot{x}_{\sigma(i)} = 0, \ |x_{\sigma(i)}| = 1, \ \text{and} \ \left(\sum_{j=1}^n a_{\sigma(i)j} x_j\right) x_{\sigma(i)} > 0, i = m+1, \cdots, \acute{n}. \qquad (3.7)$$

Let $x_I = (x_{\sigma(1)}, \cdots, x_{\sigma(m)})^T$, $x_{II} = (x_{\sigma(m+1)}, \cdots, x_{\sigma(n)})^T$,

$$A_{I,I} = [a_{\sigma(i)\sigma(j)}]_{1 \le i \le m, \ 1 \le j \le m},$$
$$A_{I,II} = [a_{\sigma(i)\sigma(j)}]_{1 \le i \le m, \ m < j \le n},$$
$$A_{II,I} = [a_{\sigma(i)\sigma(j)}]_{m < i \le n, \ 1 \le j \le m},$$
$$A_{II,II} = [a_{\sigma(i)\sigma(j)}]_{m < i \le n, \ m < j \le n}. \qquad (3.8)$$

Utilizing (3.6) through (3.9), we can represent system (1.1) equivalently as

$$\dot{x}_I = A_{I,I} x_I + A_{I,II} x_{II}$$
$$\dot{x}_{II} = 0. \qquad (3.9)$$

If we define the matrix $S = [s_{ij}]$ as $s_{ij} = 1$ if $j = \sigma(i)$ and $s_{ij} = 0$ otherwise, then it is easily verified that

$$SS^T = I, \quad x = S\begin{bmatrix} x_I \\ x_{II} \end{bmatrix}, \quad \text{and} \quad S^T AS = \begin{bmatrix} A_{I,I} & A_{I,II} \\ A_{II,I} & A_{II,II} \end{bmatrix}. \qquad (3.10)$$

We now evaluate the derivative of V with respect to t along the solutions of (1.1) as

$$\dot{V}(x(t)) = 2x(t)^T P\dot{x}(t) = 2x(t)^T Ph\Big(Ax(t)\Big)$$

$$= 2x(t)^T P Ax(t) - 2x(t)^T P\Big[Ax(t) - h\Big(Ax(t)\Big)\Big]$$

$$= -x(t)^T Qx(t) - 2[x_I(t)^T, x_{II}(t)^T]S^T P S \begin{pmatrix} 0 \\ A_{II,I}x_I(t) + A_{II,II}x_{II}(t) \end{pmatrix}$$

$$= -x(t)^T Qx(t)$$

$$-2\sum_{i=m+1}^{n}\left[\Big(p_{\sigma(i)\sigma(i)}x_{\sigma(i)}(t) + \sum_{j=1, j\neq i}^{n} p_{\sigma(j)\sigma(i)}x_j(t)\Big)\Big(\sum_{j=1}^{n} a_{\sigma(i)j}x_j(t)\Big)\right].$$

$$(3.11)$$

Noticing that $|x_{\sigma(i)}| = 1$, for all $m < i \leq n$, we obtain

$$\left(p_{\sigma(i)\sigma(i)}x_{\sigma(i)}(t) + \sum_{j=1, j\neq i}^{n} p_{\sigma(j)\sigma(i)}x_j(t)\right)\left(\sum_{j=1}^{n} a_{\sigma(i)j}x_j(t)\right)$$

$$= \left(p_{\sigma(i)\sigma(i)} + \sum_{j=1, j\neq i}^{n} p_{\sigma(j)\sigma(i)}x_j(t)x_{\sigma(i)}(t)\right)$$

$$\times \left(\sum_{j=1}^{n} a_{\sigma(i)j}x_j(t)\right)x_{\sigma(i)}(t)$$

$$\geq \left(p_{\sigma(i)\sigma(i)} - \sum_{j=1, j\neq i}^{n} |p_{\sigma(j)\sigma(i)}|\right)\left(\sum_{j=1}^{n} a_{\sigma(i)j}x_j(t)\right)x_{\sigma(i)}(t)$$

$$\geq 0, \qquad \text{for all } m < i \leq n. \qquad (3.12)$$

In the last two steps we have made use of (3.1) and (3.7) and the fact that $p_{\sigma(j)\sigma(i)}, j = 1, \cdots, n$ is a rearrangement of $p_{j\sigma(i)}, j = 1, \cdots, n$. It now follows that

$$x(t)^T P\Big[Ax(t) - h\Big(Ax(t)\Big)\Big] = \sum_{i=m+1}^{n}\left[\Big(p_{\sigma(i)\sigma(i)}x_{\sigma(i)}\right.$$

$$+ \sum_{j=1, j\neq i}^{n} p_{\sigma(j)\sigma(i)}x_j(t)\Big)\Big(\sum_{j=1}^{n} a_{\sigma(i)j}x_j(t)\Big)\right] > 0 \qquad (3.13)$$

and therefore, in view of (3.11) and (3.13), we have that

$$\dot{V}(x(t)) \leq -x(t)^T Qx(t) = \left[\frac{\partial V}{\partial x}x(t)\right]^T Ax(t).$$

This concludes the proof of the lemma. $\qquad\qquad\qquad\qquad\qquad \square$

We are now in a position to establish the next result.

Theorem 3.2.   *If there exists a positive definite, diagonally dominant matrix $P \in R^{n \times n}$ such that*

$$PA_{I'J'}^{IJ} + (A_{I'J'}^{IJ})^T P$$

is negative definite for all $(I, J, I', J') \in \Omega$ and all $A_{I'J'}^{IJ}$ defined in (3.3), then the equilibrium $x_e = 0$ of system (1.1) is globally asymptotically stable.

Proof: Let $V(x) = x^T P x$ for $x \in D^n$. Since P is positive definite and diagonally dominant, it follows from Lemma 3.1 that

$$\left[\frac{\partial V}{\partial x}(x)\right]^T h(Ax) \leq \left[\frac{\partial V}{\partial x}(x)\right]^T Ax. \tag{3.14}$$

By Lemma 3.2, it is true that

$$Ax = \sum_{(I,J,I',J')\in\Omega} \alpha_{I,J,I',J'} A_{I'J'}^{IJ} x, \quad \alpha_{I,J,I',J'} \geq 0 \quad \text{and} \quad \sum_{(I,J,I',J')\in\Omega} \alpha_{I,J,I',J'} = 1.$$

Thus,

$$\begin{aligned}
\left[\tfrac{\partial V}{\partial x}(x)\right]^T Ax &= 2x^T P Ax \\
&= 2x^T P \sum_{(I,J,I',J')\in\Omega} \alpha_{I,J,I',J'} A_{I'J'}^{IJ} x \\
&\leq max_{(I,J,I',J')\in\Omega} x^T P A_{I'J'}^{IJ} x \\
&= max_{(I,J,I',J')\in\Omega} x^T (P A_{I'J'}^{IJ} + (A_{I'J'}^{IJ})^T P) x.
\end{aligned}$$

By assumption, $P A_{I'J'}^{IJ} + (A_{I'J'}^{IJ})^T P$ is negative definite for all $(I, J, I', J') \in \Omega$. Therefore, $\left[\frac{\partial V}{\partial x}(x)\right]^T Ax$ is negative definite for all $x \in D^n$ and all $A \in [A^m, A^M]$. Applying Theorem 3.1, we conclude that the equilibrium $x_e = 0$ of system (1.1) is globally asymptotically stable. $\square$

Theorem 3.1 is in general less restrictive than Theorem 3.2; however, Theorem 3.1 is considerably easier to apply than Theorem 3.1 since there are only finitely distinct matrices $A_{I'J'}^{IJ}$ (see [2]). To verify the conditions in Theorem 3.2, we need to check the negative definiteness of 2^{2n-1} different matrices.

4. Systems with Partial State Saturation Constraints

In the present section, we establish sufficient conditions for the *global asymptotic stability* of the equilibrium $0 = z_e = (x_e^T, y_e^T)^T = (0^T, 0^T)^T$ of system (1.4). Similarly as in Section II, in the present context "$z_e = 0$ is globally asymptotically stable" means that the equilibrium $z_e = 0$ of system (1.4) is asymptotically stable and the domain of attraction is all of $R^{(n-m)} \times D^m$ (rather than R^n, since system (1.4) is defined only on $R^{(n-m)} \times D^m$).

The following result [4] is required in establishing the next result.

Lemma 4.1. *Suppose that $A = A_0$ is Hurwitz stable (i.e., all eigenvalues of A have negative real parts) and there exist symmetric positive definite matrices $P_1 \in R^{(n-m)\times(n-m)}$, $P_2 \in R^{m\times m}$, such that P_2 is diagonally dominant and such that*

$$\begin{bmatrix} P_1 & 0 \\ 0 & P_2 \end{bmatrix} A + A^T \begin{bmatrix} P_1 & 0 \\ 0 & P_2 \end{bmatrix} \tag{4.1}$$

is negative definite. Then the equilibrium $z_e = (x_e^T, y_e^T)^T = (0^T, 0^T)^T$ of system (1.4) with $A = A_0$ is globally asymptotically stable.

Proof: To prove the lemma, we choose a Lyapunov function $V : R^{(n-m)} \times D^m \to R$ of the form

$$V(z) = V\left(\begin{pmatrix} x \\ y \end{pmatrix}\right) = (x^T, y^T) \begin{bmatrix} P_1 & 0 \\ 0 & P_2 \end{bmatrix} \begin{pmatrix} x \\ y \end{pmatrix}. \tag{4.2}$$

The right-hand derivative of $V(z(t))$ along the solutions of system (1.4) is given by

$$\begin{aligned}
\dot{V}(z(t)) &= 2(x(t)^T, y(t)^T) \begin{bmatrix} P_1 & 0 \\ 0 & P_2 \end{bmatrix} \begin{pmatrix} \dot{x}(t) \\ \dot{y}(t) \end{pmatrix} \\
&= 2(x(t)^T, y(t)^T) \begin{bmatrix} P_1 & 0 \\ 0 & P_2 \end{bmatrix} \left(\begin{bmatrix} A_{11} & A_{12} \\ A_{21} & A_{22} \end{bmatrix} \begin{pmatrix} x(t) \\ y(t) \end{pmatrix} \right. \\
&\qquad \left. - \begin{pmatrix} 0 \\ A_{21}x(t) + A_{22}y(t) - h\Big(A_{21}x(t) + A_{22}y(t)\Big) \end{pmatrix} \right) \\
&= -(x(t)^T, y(t)^T)Q\begin{pmatrix} x(t) \\ y(t) \end{pmatrix} - 2y(t)^T P_2 \Big[A_{21}x(t) \\
&\qquad + A_{22}y(t) - h\Big(A_{21}x(t) + y(t)\Big) \Big].
\end{aligned} \tag{4.3}$$

By assumption, P_2 satisfies (3.1) (with $n = m$) and $y \in D^m$, and the non-zero entries of $A_{21}x(t) + A_{22}y(t) - h\Big(A_{21}x(t) + A_{22}y(t)\Big)$ have the same sign as the corresponding entries of $y(t)$. Using an identical argument as in the derivation of (3.12) and (3.13), we obtain $y(t)^T P_2 \Big[A_{21}x(t) + A_{22}y(t) - h\Big(A_{21}x(t) + A_{22}y(t)\Big) \Big] > 0$. Hence,

$$\dot{V}\left(\begin{pmatrix} x(t) \\ y(t) \end{pmatrix}\right) < -(x(t)^T, y(t)^T)Q\begin{pmatrix} x(t) \\ y(t) \end{pmatrix}. \tag{4.4}$$

We have shown that along the solutions of system (1.4) $\dot{V}$ is negative definite for all $(x^T, y^T)^T \in R^{(n-m)} \times D^m$. In view of the assumptions on P_1 and P_2, it follows that V is also positive definite and radially unbounded on $R^{(n-m)} \times D^m$. Accordingly, there exist $\phi_i : R^+ = [0, \infty) \to R^+, i = 1, 2, 3$ with $\phi_1, \phi_2 \in KR$ and $\phi_3 \in K$, such that

$$\phi_1(\|z\|) \le V(z) \le \phi_2(\|z\|)$$

and

$$\dot{V}(z) \le -\phi_3(\|z\|)$$

for all $z = (x^T, y^T)^T \in R^{(n-m)} \times D^m$ ($\phi \in KR$ if $\phi \in K$ and $\lim_{r \to \infty} \phi(r) = \infty$). Now let

$$S_c = \{z = (x^T, y^T)^T \in R^{(n-m)} \times D^m : V(z) \le c\}$$

and note that for *any solution* $z(t, z_0, t_0 = 0) = z(t) = (x(t)^T, y(t))^T$ with $z(0) = z_0$, there exists a $c > 0$ such that $z_0 \in S_c$. To conclude the proof, we now proceed in an identical manner as in the proof of Theorem 3.1 to show that the equilibrium $z_e = 0$ is globally asymptotically stable. In doing so, the compact set S_c assumes the same role as the compact set D^m did in the proof of Theorem 3.1 (recalling, that the proof of Lemma 3.1 did not depend on the particular form of the compact set D^m). We omit the details in the interests of brevity. $\square$

Theorem 4.1.   *The equilibrium $z_e = (x_e^T, y_e^T)^T = (0^T, 0^T)^T$ of system (1.4) with $A \in [A^m, A^M]$ is globally asymptotically stable if there exist symmetric positive definite matrices $P_1 \in R^{(n-m) \times (n-m)}$, $P_2 \in R^{m \times m}$, such that*

$$\begin{bmatrix} P_1 & 0 \\ 0 & P_2 \end{bmatrix} A_{I'J'}^{IJ} + (A_{I'J'}^{IJ})^T \begin{bmatrix} P_1 & 0 \\ 0 & P_2 \end{bmatrix} \tag{4.5}$$

is negative definite for all $(I, J, I', J') \in \Omega$ and all $A_{I'J'}^{IJ}$ defined in (3.3).

Proof: To prove the theorem, we choose a Lyapunov function $V : R^{(n-m)} \times D^m \to R$ of the form

$$V(z) = V\left(\begin{pmatrix} x \\ y \end{pmatrix}\right) = (x^T, y^T) \begin{bmatrix} P_1 & 0 \\ 0 & P_2 \end{bmatrix} \begin{pmatrix} x \\ y \end{pmatrix}. \tag{4.6}$$

Then the right-hand derivative of $V(z(t))$ along the solutions of system (1.4) satisfies

$$\dot{V}\left(\begin{pmatrix} x(t) \\ y(t) \end{pmatrix}\right) = (x(t)^T, y(t)^T) \begin{bmatrix} P_1 & 0 \\ 0 & P_2 \end{bmatrix} \begin{pmatrix} A_{11}x + A_{12}y \\ h(A_{21}x + A_{22}y) \end{pmatrix}$$

$$\le z(t)^T \begin{bmatrix} P_1 & 0 \\ 0 & P_2 \end{bmatrix} Az(t). \tag{4.7}$$

Using an identical argument as in the proof of Theorem 3.2, we obtain that along the solutions of system (1.4) $\dot{V}$ is negative definite for all

$(x^T, y^T)^T \in R^{(n-m)} \times D^m$ and all $A \in [A^m, A^M]$. Therefore, the equilibrium $z_e = (x_e^T, y_e^T)^T = (0^T, 0^T)^T$ of system (1.4) with $A \in [A^m, A^M]$ is globally asymptotically stable. This concludes the proof.

□

5. Examples

Example 1.   For second-order uncertain systems given by (1.1), we can give necessary and sufficient conditions for global asymptotic stability of the origin. This is based on the following result established in [4].

Lemma 5.1.   *The equilibrium $x_e = 0$ of system (1.1) with*

$$A = \begin{bmatrix} a & b \\ c & d \end{bmatrix} \in R^{2 \times 2}$$

is globally asymptotically stable if and only if A is Hurwitz stable and one of the following conditions is satisfied:

(i) $a < 0$, $d < 0$;
(ii) $a \geq |b| > 0$, $d < 0$;
(iii) $d \geq |c| > 0$, $a < 0$.

□

It follows now from Theorem 3.1 that the equilibrium $x_e = 0$ of the uncertain system given by (1.1) with $A \in [A^m, A^M] \in R^{2 \times 2}$ is globally asymptotically stable if and only if the conditions of Lemma 5.1 are true for all A. This is easily verified. For example,

(a) let

$$A^m = \begin{bmatrix} -2 & 4 \\ -2 & -3 \end{bmatrix}, \quad A^M = \begin{bmatrix} -1 & 8 \\ -1 & -2 \end{bmatrix}.$$

Since both the diagonal entries are negative, and A is Hurwitz stable for all $A \in [A^m, A^M]$, we conclude that $x_e = 0$ is globally asymptotically stable in this case.

(b) Let

$$A^m = \begin{bmatrix} -1 & -8 \\ 2 & 3 \end{bmatrix}, \quad A^M = \begin{bmatrix} -0.5 & -4 \\ 4 & 6 \end{bmatrix}.$$

The condition (iii) in Lemma 5.1 is not satisfied for

$$A = \begin{bmatrix} -1 & -4 \\ 3.5 & 3 \end{bmatrix}.$$

Therefore, $x_e = 0$ is not globally asymptotically stable.

□

Example 2. We consider system (1.1) with $A \in [A^m, A^M] \in R^{3 \times 3}$ and

$$A_m = \begin{bmatrix} -3 & -0.1 & 0.2 \\ -0.4 & -3 & 0.5 \\ 0.7 & -0.8 & -3 \end{bmatrix}, \quad A_M = \begin{bmatrix} -2 & 0.1 & 0.5 \\ 0.2 & -2 & 1.2 \\ 1.0 & 0.3 & -2 \end{bmatrix}.$$

We choose

$$P = \begin{bmatrix} 3 & 0 & 0 \\ 0 & 4 & 0 \\ 0 & 0 & 2 \end{bmatrix}.$$

P obviously satisfies the diagonal dominance condition given by (3.1). Since $n = 3$ in the present case, there are $2^{2n-1} = 32$ matrices to be checked. We used MATLAB to generate the 32 matrices $A_{I'J'}^{IJ}$ from A_m to A^M and to verify the positive definiteness of the matrices $PA_{I'J'}^{IJ} + (A_{I'J'}^{IJ})^T P$. In particular, the matrix A^M has an eigenvalue $\lambda_m(PA_{I'J'}^{IJ} + (A_{I'J'}^{IJ})^T P) = -3.6966$ which turns out to be the smallest eigenvalue among all the $PA_{I'J'}^{IJ} + (A_{I'J'}^{IJ})^T P$ involved. Thus, with the matrix P chosen as above, all conditions of Theorem 3.1 are satisfied and therefore, the equilibrium $x_e = 0$ of (1.1) with A^m and A^M given above is globally asymptotically stable.

The matrix P can be determined, using linear programming (see, e.g., [2]). □

6. Concluding Remarks

We established new sufficient conditions for the global asymptotic stability of the trivial solution of uncertain systems described by ordinary differential equations under complete saturation constraints (where all states are subject to saturation constraints) and systems with partial saturation constraints (where some of the states are subject to saturation constraints). We demonstrated the applicability of these results by means of two specific examples.

The present results constitute *robust stability* results for systems with state saturation nonlinearities. Systems of the type considered herein arise in a variety of areas, including signal processing, artificial neural networks, and control theory.

References

1. R.K. Miller and A.N. Michel, *Ordinary Differential Equations*, Academic Press, New York, 1982.

2. A. N. Michel, D. Liu and K. Wang, "Stability Analysis of a Class of Systems with Parameter Uncertainties and with State Saturation Nonlinearities", *International Journal of Robust and Nonlinear Control*, Vol. 5, pp.505-519, 1995.
3. D. Liu and A. N. Michel, "Stability Analysis of Systems with Partial State Saturation Nonlinearities", *Proc. 33rd IEEE CDC*, pp. 1311–1316, Lake Buena Vista, FL, Dec. 1994.
4. L. Hou and A. N. Michel, "Asymptotic Stability of Systems with Saturation Constraints", accepted by the *1996 IEEE CDC*.

Chapter 4. Multi-objective Bounded Control of Uncertain Nonlinear Systems: an Inverted Pendulum Example

Stéphane Dussy and Laurent El Ghaoui

Laboratoire de Mathématiques Appliquées, Ecole Nationale Supérieure de Techniques Avancées, 32, Blvd. Victor, 75739 Paris. France.

1. Introduction

1.1 Problem statement

We consider a parameter-dependent nonlinear system of the form

$$
\begin{aligned}
\dot{x} &= A(x,\xi(t))x + B_u(x,\xi(t))u + B_w(x,\xi(t))w, \\
y &= C_y(x,\xi(t))x + D_{yw}(x,\xi(t))w, \\
z &= C_z(x,\xi(t))x + D_{zu}(x,\xi(t))u, \\
\varsigma_i &= E_i x, \quad i = 1,\dots,N.
\end{aligned}
\tag{1.1}
$$

where x is the state, u the command input, y the measured output, z the controlled output, ς_i, $i = 1,\dots N$, some outputs, and w is the disturbance. The vector ξ contains time-varying parameters, which are known to belong to some given bounded set $\mathcal{D}$. Also, we assume that $A(\cdot,\cdot)$, $B_u(\cdot,\cdot)$, $C_y(\cdot,\cdot)$, $B_w(\cdot,\cdot)$, $D_{yw}(\cdot,\cdot)$ and $D_{zu}(\cdot,\cdot)$ are rational functions of their arguments. We seek a dynamic, possibly nonlinear, output-feedback controller, with input y and output u, such that a number of specifications are satisfied. To define our specifications, we consider a given polytope $\mathcal{P}$ of initial conditions, and a set of admissible disturbances, chosen to be of the form

$$
\mathcal{W}(w_{\max}) = \left\{ w \in \mathcal{L}_2([0\ \infty[) \ \middle|\ \int_0^\infty w(t)^T w(t)\,dt \le w_{\max}^2 \right\},
\tag{1.2}
$$

where $w_{\max}$ is a given scalar. To $\mathcal{P}$, $\mathcal{W}$ and the uncertainty set $\mathcal{D}$, we associate the family $\mathcal{X}(x_0, w_{\max})$ of trajectories of the above uncertain system, in response to a given $x_0 \in \mathcal{P}$. Our design constraints are as follows.

S.1 The system is well-posed along every trajectory in $\mathcal{X}(x_0, w_{\max})$, that is, for every $t \ge 0$, the system matrices $A(x,\xi(t))$, $B_u(x,\xi(t))$, etc, are well-defined.

S.2 Every trajectory in $\mathcal{X}(x_0,0)$ decays to zero at rate α, that is $\lim_{t\to\infty} e^{\alpha t} x(t) = 0$.

S.3 For every trajectory in $\mathcal{X}(x_0, w_{\max})$, the command input u satisfies $\forall t \ge 0, \|u(t)\|_2 \le u_{\max}$.

S.4 For every trajectory in $\mathcal{X}(x_0, w_{\max})$, the outputs ς_i, $i = 1, \ldots N$, are bounded, that is, $\|\varsigma_i(t)\|_2 \leq \varsigma_{i,\max}$ for every $t \geq 0$.

S.5 For every trajectory in $\mathcal{X}(0, w_{\max})$, the closed-loop system exhibits good disturbance rejection, *ie* a $\mathcal{L}_2$-gain bound γ from z to w must hold, that is for every $w(t) \in \mathcal{W}(w_{\max})$:

$$\int_0^\infty z(t)^T z(t) dt \leq \gamma^2 \int_0^\infty w^T(t) w(t) dt.$$

We emphasize that the above specifications should hold robustly, that is, for every $\xi(t) \in \mathcal{D}$.

The following sections describe an LMI-based controller synthesis method for the above problem, that is based on the results of [12]. We illustrate this method with an inverted pendulum example. The method seeks a controller that is dependent on the measured parameters and/or nonlinearities, and which yields a closed-loop system robust with respect to the unmeasured parameters and/or nonlinearities. This kind of control law can be termed "robust measurement-scheduled". It is based on a Linear Fractional Representation (LFR) of the system (see [17, 6]). The LFR model can be constructed systematically, starting from the system's dynamic equations. The LFR results in a model where the system is described by a "nominal" linear time-invariant system connected to an uncertainty diagonal matrix via (possibly fictitious) input p and output q. This matrix depends on the nonlinearities and/or uncertainties (grouped here in the vector $\xi(t)$).

The LMI approach to the gain-scheduling methodology was first introduced by Packard in [16, 3] for Linear Parameter-Varying (LPV) systems, and applied to $\mathbf{H}_\infty$-control in [18]. The extension to nonlinear systems can be found in [12]. Design examples (for LPV systems) can be found in [1, 4]. In [8], we have applied the method to a simplified inverted pendulum example. Here, we consider a more complicated system with more stringent conditions and improved performances.

1.2 Main features

There are several interesting features in the above approach.

- *Efficient numerical solution.* LMI optimization problems can be solved very efficiently using recent interior-point methods [15, 19, 10] (the global optimum is found in modest computing time). This brings a numerical solution to problems when no analytical or closed-form solution is known.
- *A systematic method.* The approach is very systematic and can be applied to a very wide variety of nonlinear control design problems. This includes (and is not reduced to) systems whose state-space representation are rational functions of the state vector.
- *Extension to uncertain nonlinear systems.* It is possible to extend the method to cases when some parameters in the state-space representation of the system are uncertain.

– *Multiobjective problems.* The approach is particularly well-suited to problems where several (possibly conflicting) specifications are to be satisfied by the closed-loop system. This is made possible by use of a common Lyapunov function *proving* that every specification holds.

In view of the above features, the approach opens the way to CAD tools for the analysis and control of nonlinear systems, *eg* the public domain toolbox MRCT [9, 7] built on top of the software `lmitool` [10].

The method is based on using the same Lyapunov function to enforce (possibly competing) design specifications. This may lead to conservative results. This disadvantage is traded off in several ways.

First, other methods, such as $\mathbf{H}_\infty$ control, do not allow to impose *directly* several constraints. One must form a single criterion, in the form of an $\mathbf{H}_\infty$-norm bound, that is deemed to reflect all specifications. Forming this one criterion is sometimes difficult, since a (sometimes large) number of parameters have to be chosen, and sometimes these parameters have no obvious relationship with the desired constraints.

In an LMI-based design, we form a set of LMI constraints, each one reflecting one specification. In each constraint appears a parameter which determines how coercive the specification is—increasing the parameter amounts to relax the constraint. For example, if we impose a bound $u_{\max}$ on the command input norm, then $u_{\max}$ is a measure of how stringent the norm bound is. The various parameters (command input $u_{\max}$, $\mathbf{L}_2$ gain bound γ, etc) can be interpreted as *design parameters.*

The approach thus reduces the synthesis problem to choosing a few design parameters, in order to meet the specifications. Note that all the design parameters have a direct physical interpretation, and that there are only a small number of them.

The possible conservatism of the LMI approach can be reduced if we choose the design parameters iteratively, as follows. At the first step of the design process, we may set the various parameters to higher values than those imposed by the actual specifications. In a second step, we may perform an (LMI-based) *analysis*, or a simulation, on the closed-loop system, in order to check if the latter obeys the specifications. (Since LMI-based design tools are more conservative than LMI-based analysis tools, it seems that this will be often the case. If not, we can identify (using our analysis or simulation) which specifications are violated, and decrease the corresponding design parameters.

1.3 Paper outline

We first detail the inverted pendulum problem in §2. and write an LFR for the system in §3.. In §4., LMI-based conditions ensuring the feasibility of the bounded multi-objective robust control problem are derived. Finally, §5. contains numerical results and simulations.

1.4 Notations

or a real matrix P, $P > 0$ (resp. $P \geq 0$) means P is symmetric and positive-definite (resp. positive semi-definite) ; then, $P^{1/2}$ denotes its symmetric square root. $\mathbf{diag}(A, B)$ denotes the block-diagonal matrix written with A, B as its diagonal blocks. For any matrix $A \in \mathbf{R}^{p \times q}$, the matrix $\mathcal{N}_A$ denotes any matrix Z such that $A^T Z = 0$ and $\mathbf{rank}[A\ Z]$ is maximum. For $r = [r_1, \ldots, r_n]$, $r_i \in \mathbf{N}$, we define the sets For $r = [r_1, \ldots, r_n]$, $r_i \in \mathbf{N}$, we define the sets

$$
\begin{aligned}
\mathcal{D}(r) &= \{\Delta = \mathbf{diag}(\delta_1 I_{r_1}, \ldots, \delta_n I_{r_n}),\ |\ \delta_i \in \mathbf{R}\}, \\
\mathcal{B}(r) &= \{B = \mathbf{diag}(B_1, \ldots, B_n),\ |\ B_i \in \mathbf{R}^{r_i \times r_i}\}, \\
\mathcal{S}(r) &= \{S \in \mathcal{B}(r)\ |\ S_i > 0,\ i = 1, \ldots, n\}.
\end{aligned}
\tag{1.3}
$$

Finally, the symbol $\mathbf{Co}\{v_1, \ldots, v_L\}$ denotes the polytope with vertices v_1, ..., v_L, and for $P > 0$ and a positive scalar λ, $\mathcal{E}_{P,\lambda}$ denotes the ellipsoid $\mathcal{E}_{P,\lambda} = \{x \mid x^T P x \leq \lambda\}$.

2. The inverted pendulum example

The inverted pendulum is built on a cart moving in translation without constraint (see figure 2.1). Our goal is to stabilize the it in the vertical position, with an uncertain but constant weight m.

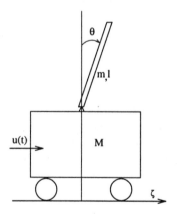

Fig. 2.1. Inverted pendulum on a cart

In fig.2.1, ζ is the position of the cart, θ is the angle position of the pendulum with the vertical axis, u is the control input acting on the cart translation and w is the disturbance. We seek to control the angle and the translation position, θ and ζ. The nonlinear dynamic equations of the system are:

$$
\begin{cases}
[m\cos^2\theta - \frac{4}{3}(M+m)]\ddot{\theta} &= m\cos\theta\sin\theta\dot{\theta}^2 - \frac{q}{l}(M+m)\sin\theta \\
& \quad + \frac{1}{l}\cos\theta u(t) + \frac{1}{l}w(t), \\
[m\cos^2\theta - \frac{4}{3}(M+m)]\ddot{\zeta} &= mg\sin\theta\cos\theta - \frac{4}{3}ml\dot{\theta}^2\sin\theta - \frac{4}{3}u(t) \\
& \quad + \kappa w(t), \\
y = \begin{bmatrix} \theta & \zeta \end{bmatrix}^T, & \\
z = \begin{bmatrix} \sqrt{0.1}\theta & \sqrt{0.1}\zeta & u \end{bmatrix}^T. &
\end{cases}
$$

$$(2.1)$$

We assume that the mass m is constant and unknown-but-bounded: $m \in [m_0 - \Delta m, \ m_0 + \Delta m]$, where m_0, Δm are given. The parameter vector here is taken to be $\xi = (m - m_0)/\Delta m$, so that $\xi \in \mathcal{D} = [-1, \ 1]$. Our specifications are described by **S.1-5**, with α, $u_{\max}$, γ and $z_{\max} = [\theta_{max} \ \dot{\theta}_{max}]^T$ given . The polytope of admissible initial conditions is determined by two given scalars θ_0, ζ_0: $\mathcal{P} = \mathbf{Co}\{v_+, -v_+, v_-, -v_-\}$, with $v_\pm = [\pm\theta_0 \ \pm\zeta_0]^T$. Moreover, the system is also perturbed by a disturbance $w(t) \in \mathcal{L}_2([0, \infty[)$.

We seek a control law that stabilizes both the angle position and the translation motion of the cart around a given equilibrium position.

3. Construction of the LFR

3.1 LFR of the open-loop system

The first step is to derive a Linear-Fractional Representation (LFR) for the system. In other words, we seek a set of equations

$$
\begin{aligned}
\dot{x} &= Ax + B_p p + B_u u + B_w w, \\
q &= C_q x + D_{qp} p + D_{qu} u + D_{qw} w, \\
y &= C_y x + D_{yp} p + D_{yw} w, \\
z &= C_z x + D_{zp} p + D_{zu} u, \\
p &= \Delta(x, \xi(t))q, \quad \Delta \in \mathcal{D}(r).
\end{aligned}
$$

$$(3.1)$$

that is *formally* equivalent to the nonlinear equations (2.1). We will refer to [8] where the framework for the construction and the normalization of an LFR for such a system is described.

Introducing $\delta_1 = \cos\theta$, $\delta_2 = \frac{\sin\theta}{\theta}$, $\delta_3 = \dot{\theta}\sin\theta$, $\delta_4 = m \in [m_0 - \Delta m, \ m_0 + \Delta m]$, we may rewrite (2.1) in

$$
\begin{bmatrix} \dot\theta \\ \ddot\theta \\ \dot\zeta \\ \ddot\zeta \end{bmatrix} = \begin{bmatrix} 0 & 1 & 0 & 0 \\ -\dfrac{3g(M+\delta_4)\delta_2}{3l\delta_4\delta_1^2-4(M+\delta_4)l} & \dfrac{3l\delta_4\delta_1\delta_3}{3l\delta_4\delta_1^2-4(M+\delta_4)l} & 0 & 0 \\ 0 & 0 & 0 & 1 \\ \dfrac{3gl\delta_4\delta_1\delta_2}{3l\delta_4\delta_1^2-4(M+\delta_4)l} & \dfrac{-4l^2\delta_4\delta_3}{3l\delta_4\delta_1^2-4(M+\delta_4)l} & 0 & 0 \end{bmatrix} \begin{bmatrix} \theta \\ \dot\theta \\ \zeta \\ \dot\zeta \end{bmatrix}
$$
$$
+ \begin{bmatrix} 0 \\ \dfrac{3\delta_1}{3l\delta_4\delta_1^2-4(M+\delta_4)l} \\ 0 \\ \dfrac{-4l}{3l\delta_4\delta_1^2-4(M+\delta_4)l} \end{bmatrix} u(t) + \begin{bmatrix} 0 \\ \dfrac{3}{3l\delta_4\delta_1^2-4(M+\delta_4)l} \\ 0 \\ \dfrac{3l}{3l\delta_4\delta_1^2-4(M+\delta_4)l} \end{bmatrix} w(t). \tag{3.2}
$$

The coefficients of the above state matrices are rational functions of the parameters δ_i, $i = 1,\dots 4$, so that we may derive an LFR for the system. This LFR has the form (3.1) with A, B_p, ..., given in Appendix A.1 and with $\Delta = \mathbf{diag}(\Delta_m, \Delta_u) = \mathbf{diag}(\delta_1 I_6, \delta_2, \delta_3, \delta_4 I_4)$ where $\Delta_m = \mathbf{diag}(\delta_1 I_6, \delta_2)$ is measured in real time, while $\Delta_u = \mathbf{diag}(\delta_3, \delta_4 I_4)$ is not.

3.2 LFR of the controller

We take our measurement-scheduled controller to be in the LFR format

$$
\begin{aligned}
\dot{\bar{x}} &= \overline{A}\bar{x} + \overline{B}_y y + \overline{B}_p \bar{p}, \quad \bar{x}(0) = 0, \\
\bar{q} &= \overline{C}_q \bar{x} + \overline{D}_{qy} y + \overline{D}_{qp} \bar{p}, \\
u &= \overline{C}_u \bar{x}, \\
\bar{p} &= \Delta_m(y)\bar{q}.
\end{aligned} \tag{3.3}
$$

where $\bar{p}$, $\bar{q}$ are fictitious measured inputs and outputs of the controller. Note the crucial assumption here: the controller is scheduled with respect to the measured element in the nonlinear block Δ, namely $\Delta_m(y)$. In our benchmark example, this means that the controller matrices will be scheduled on the value of $\delta_1 = \cos\theta$ and $\delta_2 = \frac{\sin\theta}{\theta}$.

3.3 LFR of the closed-loop system

With the chosen controller structure, the closed-loop system assumes the form

$$
\dot{\tilde{x}} = \tilde{A}(\tilde{\Delta})\tilde{x} + \tilde{B}_w(\tilde{\Delta})w, \tag{3.4}
$$

where $\tilde{x} = [x^T\ \bar{x}^T]^T$. Here, $\tilde{A}$, $\tilde{B}_w$ are given rational functions of the nonlinearity element $\tilde{\Delta}$, where the part $\Delta_m(y)$ appears twice as it is present in the open-loop system and in the controller (see Fig. 3.1).

$$
\tilde{\Delta} = \mathbf{diag}(\Delta_u(x), \Delta_m(y), \Delta_m(y)). \tag{3.5}
$$

Introducing $\tilde{p}^T = \begin{bmatrix} p^T & \bar{p}^T \end{bmatrix}$, $\tilde{q}^T = \begin{bmatrix} q^T & \bar{q}^T \end{bmatrix}$, we obtain the following LFR for the closed-loop system:

$$\begin{aligned}
\dot{\tilde{x}} &= \tilde{A}\tilde{x} + \tilde{B}_p\tilde{p} + \tilde{B}_w w, \\
\tilde{q} &= \tilde{C}_q\tilde{x} + \tilde{D}_{qp}\tilde{p} + \tilde{D}_{qw} w, \\
z &= \tilde{C}_z\tilde{x} + \tilde{D}_{zp}\tilde{p}, \\
\tilde{p} &= \tilde{\Delta}\tilde{q}, \quad \tilde{\Delta} = \mathbf{diag}(\Delta_u(x), \Delta_m(y), \Delta_m(y)),
\end{aligned} \tag{3.6}$$

where the matrices $\tilde{A}$, $\tilde{B}_p$, etc, depend affinely in the design matrix variables $\overline{C}_u$ and K, defined by

$$K = \left[\begin{array}{ccc} \overline{A} & \overline{B}_p & \overline{B}_y \\ \overline{C}_q & \overline{D}_{qp} & \overline{D}_{qy} \end{array} \right].$$

The closed-loop system is shown in Fig. 3.1.

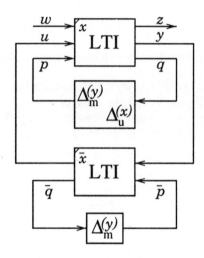

Fig. 3.1. Closed-loop system.

4. Synthesis Conditions

4.1 Analysis via quadratic Lyapunov functions

We seek quadratic Lyapunov functions ensuring specifications **S.1-5** for the closed-loop system. We thus consider a matrix $\tilde{P} \in \mathbf{R}^{n \times \tilde{n}}$, with $\tilde{P} > 0$, and associated Lyapunov function $V(\tilde{x}) = \tilde{x}^T \tilde{P}\tilde{x}$. Assume that, for a given nonzero trajectory in the set $\mathcal{X}(x_0, w_{max})$, V satisfies

$$\dot{V} + 2\alpha V + \nu(z^T z - \gamma^2 w^T w) < 0. \tag{4.1}$$

When $\nu = 0$, the above condition ensures the α-stability of the system. When $\nu \neq 0$ and $\alpha = 0$, it guarantees a maximum $\mathcal{L}_2$-gain from w to z, and when

$\nu \neq 0$ and $\alpha \neq 0$, it both ensures the α-stability of the unperturbed system and guarantees maximum $\mathcal{L}_2$-gain from w to z when the initial conditions are set to zero.

We can readily infer from (4.1) that, for a given nonzero trajectory in the set $\mathcal{X}(x_0, w_{\max})$, with $\tilde{x}_0 \in \mathcal{E}_{\tilde{P}, \lambda}$, the following condition holds

$$V(\tilde{x}(T)) < \lambda + \nu \gamma^2 w_{\max}^2. \tag{4.2}$$

The above bound implies that when initiating in an ellipsoid $\mathcal{E}_{P,\lambda}$, the augmented state of the system $\tilde{x}(t)$ will be confined in the bigger ellipsoid $\mathcal{E}_{\tilde{P},(\lambda + \nu \gamma^2 w_{\max}^2)}$. In other words, it suffices to ensure that inequality (4.1) holds to prove that every given nonzero trajectory in the set $\mathcal{X}(x_0, w_{\max})$ will never escape this bigger ellipsoid. This "pseudo-invariance condition" will be used in the following sections.

4.2 Stability and disturbance rejection conditions

As seen in [12, 8], it is possible to formulate the synthesis conditions to ensure the specifications **S.2** and **S.5** in the form of a set of LMI conditions associated with a non-convex constraint. These inequalities depend in particular on P and Q, the upper-left $n \times n$ blocks of $\tilde{P}$ and $\tilde{P}^{-1}$, respectively.

Theorem 4.1. *Let $\mathcal{N}$ be a matrix whose columns span the null-space of $[C_y \; D_{yp} \; D_{yw}]^T$. The synthesis conditions are*

$$\exists P, Q, S = \begin{bmatrix} S_m & 0 \\ 0 & S_u \end{bmatrix}, T = \begin{bmatrix} T_m & 0 \\ 0 & T_u \end{bmatrix} \in \mathcal{S}(r) \text{ and } Y \text{ such that}$$

$$\begin{bmatrix} P & I \\ I & Q \end{bmatrix} \geq 0, \quad \begin{bmatrix} S & I \\ I & T \end{bmatrix} \geq 0, \tag{4.3}$$

$$\mathcal{N}^T \begin{bmatrix} \begin{matrix} A^T P + PA + C_q^T S C_q \\ + C_z^T C_z + 2\alpha P \end{matrix} & \begin{matrix} PB_p + C_q^T S D_{qp} \\ + C_z^T D_{zp} \end{matrix} & PB_w + C_q^T S D_{qw} \\ * & \begin{matrix} D_{qp}^T S D_{qp} - S \\ + D_{zp}^T S D_{zp} \end{matrix} & D_{qp}^T S D_{qw} \\ * & * & D_{qw}^T S D_{qw} - \gamma^2 I \end{bmatrix} \mathcal{N} < 0,$$

$$\tag{4.4}$$

$$\begin{bmatrix} \begin{matrix} AQ + QA^T + B_u Y + Y^T B_u \\ + B_p T B_p^T + \gamma^{-2} B_w B_w^T + 2\alpha Q \end{matrix} & \begin{matrix} QC_q^T + Y^T D_{qu}^T \\ + B_p T D_{qp}^T + \gamma^{-2} B_w D_{qw}^T \end{matrix} & \begin{matrix} QC_z^T + Y^T D_{zu}^T \\ B_p T D_{zp}^T \end{matrix} \\ * & \begin{matrix} D_{qp} T D_{qp}^T - T \\ + \gamma^{-2} D_{qw} D_{qw}^T \end{matrix} & D_{qp} T D_{zp}^T \\ * & * & D_{zp} T D_{zp}^T - I \end{bmatrix} < 0,$$

$$\tag{4.5}$$

$$S_u T_u = I_{\nu_u}. \tag{4.6}$$

4.3 Command input bound

Theorem 4.2. *A sufficient condition to have $\|u(t)\|_2 \leq u_{\max}$ for every $t \geq 0$ is that the matrices P, Q, Y as defined in § 4.2 and variables λ, $\mu > 0$ also satisfy*

$$x_0 \in \mathcal{E}_{P,\lambda}, \tag{4.7}$$

$$\mu(\lambda + \nu\gamma^2 w_{\max}^2) = 1, \tag{4.8}$$

$$\begin{bmatrix} \mu u_{\max}^2 I & Y & 0 \\ Y^T & Q & I \\ 0 & I & P \end{bmatrix} \geq 0. \tag{4.9}$$

Proof: We defined in § 4.2 that P and Q are the upper-left $n \times n$ blocks of $\tilde{P}$ and $\tilde{P}^{-1}$. Precisely, $\tilde{P}$ and $\tilde{Q}$ are parameterized with arbitrary matrices M and L as follows (see [12] for more details)

$$\tilde{P} = \begin{bmatrix} P & M \\ M^T & \overline{P} \end{bmatrix}, \quad \tilde{Q} = \begin{bmatrix} Q & L \\ L^T & \overline{Q} \end{bmatrix},$$

where $\overline{P}$ and $\overline{Q}$ are such that $\mathcal{E}_{\overline{P}^{-1}}$ and $\mathcal{E}_{\overline{Q}^{-1}}$ are respectively the projection of the ellipsoids $\mathcal{E}_{\tilde{P}}$ and $\mathcal{E}_{\tilde{Q}}$ on the subspace of the controller state. Then, the matrix Y defined in § 4.2 also depends on the the parameterization matrices via $Y = \overline{C}_u L^T$. Let assume that the above conditions (4.7), (4.8) and (4.9) hold. Then, using Schur complement [5], (4.9) is equivalent to

$$P > 0 \text{ and } \begin{bmatrix} \mu u_{\max}^2 & Y \\ Y^T & Q - P^{-1} \end{bmatrix} \geq 0.$$

Then, with $Y = \overline{C}_u L^T$, we readily obtain

$$P > 0 \text{ and } \begin{bmatrix} \mu u_{\max}^2 & \overline{C}_u \\ \overline{C}_u^T & L^{-1}(Q - P^{-1})L^{-T} \end{bmatrix} \geq 0.$$

With $\overline{Q} = L^T(Q - P^{-1})^{-1}L$ (which stems from the matrix equality $\tilde{P}\tilde{Q} = I$), we may infer the following equivalent conditions

$$P > 0 \text{ and } \begin{bmatrix} \overline{Q}^{-1} & \overline{C}_u^T \\ \overline{C}_u & \mu u_{\max}^2 \end{bmatrix} \geq 0.$$

Using Schur complements, we obtain

$$P > 0 \text{ and } \forall \overline{x} \in \mathbf{R}^n, \; \overline{x}^T \overline{Q}^{-1} \overline{x} - \overline{x}^T \frac{\overline{C}_u^T \overline{C}_u}{\mu u_{\max}^2} \overline{x} \geq 0.$$

The assumption $x_0 \in \mathcal{E}_{P,\lambda}$ implies that $\tilde{x} \in \mathcal{E}_{\tilde{P}, \frac{1}{\mu}}$. Therefore, the state of the controller will never escape the ellipsoid obtained by projecting the state-space of the augmented system on the subspace of the controller space. In other words, that means that $\overline{x} \in \mathcal{E}_{\overline{Q}^{-1}, \frac{1}{\mu}}$. Then, with $u = \overline{C}_u \overline{x}$, we can write

$$\forall\, x_0 \in \mathcal{E}_{P,\lambda}, \quad u^T u \leq u_{\max}^2.$$

□

Remark: As seen in [5], the condition $x_0 \in \mathcal{E}_{P,\lambda}$ for every $x_0 \in$ $\mathrm{Co}\{v_1, \ldots, v_L\}$ can be written as the following LMIs

$$v_j^T P v_j \leq \lambda, \quad j = 1, \ldots L.$$

4.4 Output bound

Theorem 4.3. *A sufficient condition to have, for every $t \geq 0$, $\|\varsigma_i(t)\|_2 =$ $\|E_i x(t)\|_2 \leq \varsigma_{i,\max}$, $i = 1, \ldots N$, is that the matrix Q as defined in § 4.2 and variables λ, $\mu > 0$ also satisfies (4.7), (4.8) and*

$$\mu \varsigma_{i,\max}^2 - E_i^T Q E_i \geq 0, \quad i = 1, \ldots N. \tag{4.10}$$

Proof: Let assume that the above conditions hold. Then using Schur complements with the additional condition $Q > 0$, (4.10) is equivalent to

$$\begin{bmatrix} Q^{-1} & E_i^T \\ E_i & \mu \varsigma_{i,\max}^2 \end{bmatrix} \geq 0, \quad i = 1, \ldots N.$$

Another Schur complement transformation leads to

$$\forall\, x \in \mathbf{R}^n, \quad x^T Q^{-1} x \geq x^T \frac{E_i^T E_i}{\mu \varsigma_{i,\max}^2} x, \quad i = 1, \ldots N.$$

Now, if $x_0 \in \mathcal{E}_{P,\lambda}$, then $x \in \mathcal{E}_{P,\mu}$. Furthermore, we know from (4.3) that $P \geq Q^{-1}$, which implies that $x \in \mathcal{E}_{Q^{-1},\mu}$. This achieves the proof since then

$$\forall\, x \in \mathbf{R}^n \text{ s.t. } x_0 \in \mathcal{E}_{P,\lambda}, \quad \varsigma_{i,\max} \geq x^T E_i^T E_i x = \varsigma_i^T \varsigma_i, \quad i = 1, \ldots N.$$

□

4.5 Solving for the synthesis conditions

Every condition above is an LMI, except for the non-convex equations (4.6) and (4.8). When (4.3) holds, enforcing these conditions can be done by imposing $\mathbf{Tr} S_u T_u + \mu(\lambda + w_{\max}^2 \gamma^2) = \nu_u + 1$ with the additional constraint

$$\begin{bmatrix} \mu & I \\ I & \lambda + \gamma^2 w_{\max}^2 \end{bmatrix} \geq 0. \tag{4.11}$$

In fact, the problem belongs to the class of "cone-complementarity problems", which are based on LMI constraints of the form

$$F(V, W, Z) \geq 0, \quad \begin{bmatrix} V & I \\ I & W \end{bmatrix} \geq 0, \tag{4.12}$$

where V, W, Z are matrix variables (V, W being symmetric and of same size), and $F(V, W, Z)$ is a matrix-valued, affine function, with F symmetric. The corresponding cone-complementarity problem is

$$\text{minimize } \mathbf{Tr}VW \text{ subject to (4.12).} \tag{4.13}$$

The heuristic proposed in [11], which is based on solving a sequence of LMI problems, can be used to solve the above non-convex problem. This heuristic is guaranteed to converge to a stationary point.

Algorithm $\mathcal{H}$:

1. Find V, W, Z that satisfy the LMI constraints (4.12). If the problem is infeasible, stop. Otherwise, set $k = 1$.
2. Find V_k, W_k that solve the LMI problem

$$\text{minimize } \mathbf{Tr}(V_{k-1}W + W_{k-1}V) \text{ subject to (4.12).}$$

3. If the objective $\mathbf{Tr}(V_{k-1}W_k + W_{k-1}V_k)$ has reached a stationary point, stop. Otherwise, set $k = k+1$ and go to (2).

4.6 Reconstruction of the controller

When the algorithm exits successfully, *ie* when $S_u T_u \simeq I$ and $\mu(\lambda + w_{\max}^2 \gamma^2) \simeq 1$ (note that a stopping criterion is given in [11]), then we can reconstruct an appropriate controller by solving another LMI problem [12, 8] in the controller variables K only, C_u being directly inferred from the variables Y, P, Q. Note that analytic controller formulae are given in [13].

5. Numerical results

Our design results are based on two sets of LMIs: the first set focuses on the α-stability of the system for given initial conditions. It corresponds to the case $\nu = 0$ in § 4.1. The second one guarantees a maximum $\mathcal{L}_2$-gain for the system, as defined in §4.2, with $\alpha = 0$. (It may be too conservative to try and enforce both a decay-rate and an $\mathcal{L}_2$-gain condition via the same set of LMIs.) The bounds (respectively in $rad.s^{-1}$ and in rad) on the states that appear in the nonlinearities are $z_{\max} = [\theta_{\max} \ \dot{\theta}_{\max}]^T = [0.5 \ 1]^T$, and the system parameters are $M = 1$, $l = 0.6$, $g = 10$, $m = 0.2(1 \pm 0.4\delta_m)$ with $|\delta_m| \leq 1$. The units used for the system parameters and for the plots are kilogram for weights, meter for lengths, second for time, radian for angles and Newton for forces. The design parameters are $\alpha = 0.05$, $x_0 = [\theta_0 \ \dot{\theta}_0 \ \zeta_0 \ \dot{\zeta}_0]^T = [0 \ 0.2 \ 0 \ 0.2]^T$. We seek to minimize the bound $u_{\max}$ on the command input and to achieve good disturbance rejection. The numerical results were obtained using the public domain toolbox MRCT [9, 7] built on top of the software lmitool [10] and the Semidefinite Programming package **SP** [19]. The output-feedback

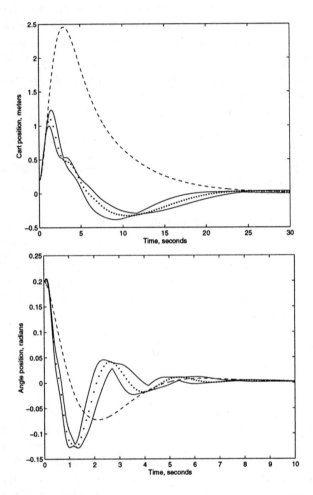

Fig. 5.1. Closed-loop system responses (cart position and angle position) with $w \equiv 0$ and nonzero initial conditions. Extremal responses (OF) in plain line, nominal response (OF) in dotted line, nominal response (SF) in dashed line.

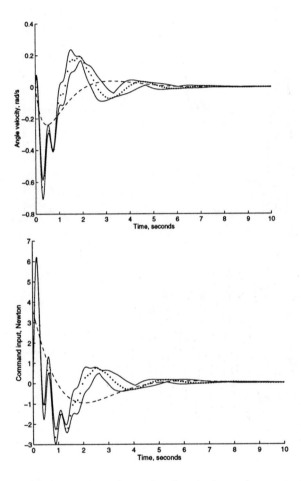

Fig. 5.2. Closed-loop system responses (angle velocity and command input) with $w \equiv 0$ and nonzero initial conditions. Extremal responses (OF) in plain line, nominal response (OF) in dotted line, nominal response (SF) in dashed line.

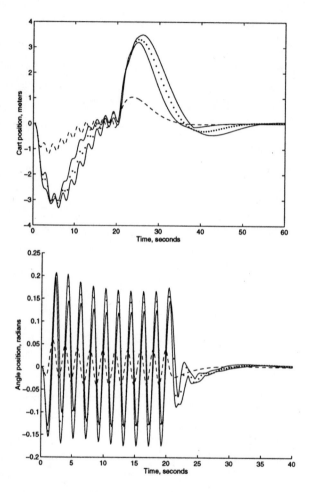

Fig. 5.3. Closed-loop system responses (cart position and angle position) with disturbance $w(t)$ and zero initial conditions. Extremal responses (OF) in plain line, nominal response (OF) in dotted line, nominal response (SF) in dashed line.

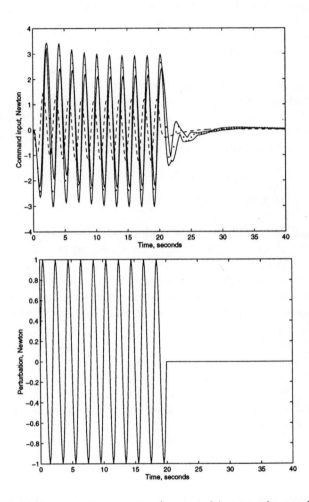

Fig. 5.4. Closed-loop system responses (command input and perturbation) with disturbance $w(t)$ and zero initial conditions. Extremal responses (OF) in plain line, nominal response (OF) in dotted line, nominal response (SF) in dashed line.

controller is given by the LFR (3.3), where the gain matrices are described in Appendix A.2.

We have shown the closed-loop system responses with $w \equiv 0$ and a nonzero initial condition $\theta_0 = 0.2$rad, $z_0 = 0.2$m in Fig. 5.1-??, and with zero initial conditions and $w(t) = 0.1\cos(\pi t)$, $0 \le t \le 20$ in Fig. ??. We plotted in plain line the envelopes of the responses for different values of the weight m varying from -40% to $+40\%$ of the nominal value. We also plotted the response of the nominal system ($m = 0.2$kg) with the output-feedback law (dotted line), and with a state-feedback control law (dashed line) $u = [17.89 \; 4.65 \; 0.06 \; 0.40]x$ computed as described in [8]. Both robust stabilization and disturbance rejection are achieved for a 40% variation of the weight around its nominal value. We can note that the bounds on $\theta(t)$ and on $\dot{\theta}(t)$, respectively $\theta_{\max} = 0.5$rad and $\dot{\theta}_{\max} = 1$rad/s, are enforced. Also note that the state-feedback controller requires less energy ($u_{\max}^{SF} = 3.8$Newton) than the output-feedback one ($u_{\max}^{OF} = 6.2$Newton).

6. Concluding remarks

In this paper, an LMI-based methodology was described to ensure various specifications of stability, disturbance rejection, input and output bounds, for a large class of uncertain, perturbed, nonlinear systems. The proposed control law is allowed to depend on parameters or states that are measured in real time, while it remains robust in respect to the unmeasured ones. This methodology has been successfully applied to the control of an uncertain, perturbed inverted pendulum. It is very systematic and handles *multiobjective* control. One important aspect of the method is that it allows computing trade-off curves for multiobjective design.

Recently, accurate robustness analysis tools based on multiplier theory have been devised, see *e.g.* Megretsky and Rantzers [14] and Balakrishnan [2]. A complete methodology for nonlinear design should include these tools for controller validation.

References

1. P. Apkarian, P. Gahinet, and G. Becker. Self-scheduled $\mathcal{H}_\infty$ control of linear parameter-varying systems: a design example. *Automatica*, 31(9):1251–1261, September 1995.
2. V. Balakrishnan. Linear Matrix Inequalities in robustness analysis with multipliers. *Syst. & Contr. Letters*, 25(4):265–272, July 1995.
3. G. Becker and A. Packard. Robust performance of linear parametrically varying systems using parametrically-dependent linear feedback. *Syst. & Contr. Letters*, 23(3):205–215, September 1994.

4. B. Bodenheimer and P. Bendotti. Optimal linear parameter-varying control design for a pressurized water reactor. In *Proc. IEEE Conf. on Decision & Contr.*, pages 182–187, New Orleans, LA, December 1995.

5. S. Boyd, L. El Ghaoui, E. Feron, and V. Balakrishnan. *Linear Matrix Inequality in systems and control theory.* SIAM, 1994.

6. J.C. Doyle, A. Packard, and K. Zhou. Review of $\mathcal{LFT}$s, LMIs and μ. In *Proc. IEEE Conf. on Decision & Contr.*, pages 1227–1232, Brignton, England, December 1991.

7. S. Dussy and L. El Ghaoui. *Multi-objective Robust Control Toolbox (MRCT): user's guide*, 1996. Available via http://www.ensta.fr/~gropco/staff/dussy/gocpage.html.

8. S. Dussy and L. El Ghaoui. Robust gain-scheduled control of a class of nonlinear parameter-dependent systems: application to an uncertain inverted pendulum. In *Proc. Conf. on Contr. & Applications*, pages 516–521, Dearborn, MI, September 1996.

9. S. Dussy and L. El Ghaoui. Multiobjective Robust Control Toolbox for LMI-based control. In *Proc. IFAC Symposium on Computer Aided Control Systems Design*, Gent, Belgium, April 1997.

10. L. El Ghaoui, R. Nikoukhah, and F. Delebecque. LMITOOL: *A front-end for LMI optimization, user's guide*, February 1995. Available via anonymous ftp to ftp.ensta.fr, under /pub/elghaoui/lmitool.

11. L. El Ghaoui, F. Oustry, and M. Ait Rami. An LMI-based linearization algorithm for static output-feedback and related problems. *IEEE Trans. Aut. Contr.*, May 1997.

12. L. El Ghaoui and G. Scorletti. Control of rational systems using Linear-Fractional Representations and Linear Matrix Inequalities. *Automatica*, 32(9):1273–1284, September 1996.

13. T. Iwasaki and R.E. Skelton. All controllers for the general $\mathcal{H}\infty$ control problems: LMI existence conditions and state space formulas. *Automatica*, 30(8):1307–1317, August 1994.

14. A. Megretski and A. Rantzer. System analysis via integral quadratic constraints. In *Proc. IEEE Conf. on Decision & Contr.*, pages 3062–3067, Orlando, FL, December 1994.

15. Y. Nesterov and A. Nemirovsky. *Interior point polynomial methods in convex programming: theory and applications.* SIAM, 1993.

16. A. Packard. Gain scheduling via Linear-Fractional Transformations. *Syst. & Contr. Letters*, 22(2):79–92, February 1994.

17. A. Packard, K. Zhou, P. Pandey, and G. Becker. A collection of robust control problems leading to LMI's. In *Proc. IEEE Conf. on Decision & Contr.*, pages 1245–1250, Brighton, England, December 1991.

18. G. Scorletti and L. El Ghaoui. Improved Linear Matrix Inequalities conditions for gain-scheduling. In *Proc. IEEE Conf. on Decision & Contr.*, pages 3626–3631, New Orleans, LA, December 1995.

19. L. Vandenberghe and S. Boyd. **SP**, *Software for semidefinite programming, user's guide*, December 1994. Available via anonymous ftp to isl.stanford.edu under /pub/boyd/semidef_prog.

A. Linear Fractional Representation

A.1 LFR of the open-loop system

The LFR of the inverted pendulum is described by (3.1) with the following matrices.

$$
A = \begin{bmatrix} 0 & 1 & 0 & 0 \\ 0 & 0 & 0 & 0 \\ 0 & 0 & 0 & 1 \\ 0 & 0 & 0 & 0 \end{bmatrix}, \quad
B_u = \begin{bmatrix} 0 \\ 0 \\ 0 \\ \frac{1}{M} \end{bmatrix}, \quad
B_w = \begin{bmatrix} \frac{-3}{4M} \\ 0 \\ \frac{-3}{4Ml} \\ 0 \end{bmatrix},
$$

$$
C_q = \begin{bmatrix} 0 & 0 & 0 & 0 \\ 0 & 0 & 0 & 0 \\ 0 & 0 & 0 & 0 \\ 0 & 0 & 0 & 0 \\ 0 & 0 & 0 & 0 \\ 0 & 0 & 0 & 0 \\ 1 & 0 & 0 & 0 \\ 0 & 1 & 0 & 0 \\ 0 & 0 & 0 & 0 \\ 0 & 0 & 0 & 0 \\ 0 & 0 & 0 & 0 \\ 0 & 0 & 0 & 0 \\ 0 & 0 & 0 & 1 \end{bmatrix}, \quad
D_{qu} = \begin{bmatrix} 0 \\ 0 \\ 0 \\ \frac{1}{l} \\ 0 \\ \frac{1}{M} \\ 0 \\ 0 \\ 0 \\ 0 \\ \frac{-4}{3M} \\ 0 \end{bmatrix}, \quad
D_{qw} = \begin{bmatrix} 0 \\ 0 \\ \frac{-3}{4M} \\ 0 \\ 0 \\ \frac{-3}{4Ml} \\ 0 \\ 0 \\ 0 \\ 0 \\ \frac{1}{Ml} \\ \frac{1}{M} \end{bmatrix},
$$

$$
B_p = \frac{3}{4M} \begin{bmatrix} 0 & 0 & 0 & 0 & 0 & 0 & 0 & 0 & 0 & 0 & 0 & 0 \\ 0 & 0 & 0 & -1 & 0 & 0 & \frac{gM}{l} & 0 & \frac{g}{l} & 0 & 0 & 1 \\ 0 & 0 & 0 & 0 & 0 & 0 & 0 & 0 & 0 & 0 & 0 & 0 \\ -1 & 0 & 0 & 0 & 0 & 0 & 0 & 0 & 0 & \frac{4}{3l} & 1 & 0 \end{bmatrix},
$$

$$
D_{qp} = \begin{bmatrix} 0 & 0 & 0 & 0 & 0 & 0 & 0 & 0 & g & 0 & 0 & 0 \\ 0 & 0 & 1 & 0 & 0 & 0 & 0 & 0 & 0 & 0 & 0 & 0 \\ 0 & 0 & 0 & \frac{-3}{4M} & 0 & 0 & \frac{3g}{4l} & 0 & \frac{3g}{4Ml} & 0 & 0 & \frac{3}{4M} \\ 0 & 0 & 0 & 0 & 0 & 0 & 0 & 0 & 0 & 1 & 0 & 0 \\ 0 & 0 & 0 & 0 & 0 & 1 & 0 & 0 & 0 & 0 & 0 & 0 \\ \frac{-3}{4M} & 0 & 0 & 0 & 0 & 0 & 0 & 0 & 0 & \frac{1}{lM} & \frac{3}{4M} & 0 \\ 0 & 0 & 0 & 0 & 0 & 0 & 0 & 0 & 0 & 0 & 0 & 0 \\ 0 & 0 & 0 & 0 & 0 & 0 & 0 & 0 & 0 & 0 & 0 & 0 \\ 0 & 0 & 0 & 0 & 0 & 0 & 1 & 0 & 0 & 0 & 0 & 0 \\ 0 & 0 & 0 & 0 & 0 & 0 & 0 & 1 & 0 & 0 & 0 & 0 \\ \frac{1}{M} & 0 & 0 & 0 & 1 & 0 & 0 & 0 & 0 & \frac{-4}{3lM} & \frac{-1}{M} & 0 \\ 0 & 1 & 0 & \frac{1}{M} & 0 & 0 & \frac{-g}{l} & 0 & \frac{-g}{Ml} & 0 & 0 & \frac{-1}{M} \end{bmatrix}.
$$

A.2 LFR of the controller

The LFR of the controller is described by (3.3) with the following gain matrices.

$$
\overline{A} = \begin{bmatrix} -8.8 & -3.7 & -1.2 & -12.7 \\ -54.2 & -19.7 & -7.6 & -13.0 \\ 39.3 & 19.0 & -24.1 & -168.3 \\ -1115 & -443 & -181 & -1062 \end{bmatrix}, \quad \overline{C}_u = 10^{-3} \begin{bmatrix} -9.9 \\ -16.0 \\ 28.9 \\ -9.6 \end{bmatrix}^T,
$$

$$
\overline{B}_y = \begin{bmatrix} -395 & 44 \\ 1159 & 280 \\ 8868 & -157 \\ -732 & 5780 \end{bmatrix}, \quad \overline{D}_{qy} = 0.1 \begin{bmatrix} -22.9 & 0.1 \\ 94.8 & -1.4 \\ -53.7 & 1.5 \\ -82.5 & 1.6 \\ -315 & 1.4 \\ 13.4 & -0.1 \\ -1.4 & 0.0 \end{bmatrix},
$$

$$
\overline{B}_p = 0.01 \begin{bmatrix} 0.1 & 0.5 & -3.2 & 24.0 & -1.7 & -1.7 & -0.0 \\ -0.4 & -2.5 & 12.5 & -53.5 & 3.3 & 3.9 & 0.1 \\ -6.8 & -22.4 & 114 & -267 & 22.7 & 21.4 & -4.0 \\ 6.3 & 20.4 & -89 & 111.3 & -10.4 & -10.6 & 2.4 \end{bmatrix},
$$

$$
\overline{C}_q = 0.01 \begin{bmatrix} -0.6 & -0.3 & 0.6 & 0.9 \\ 4.7 & 1.2 & -2.7 & -8.4 \\ -3.9 & 0.4 & -0.4 & 9.2 \\ -8.1 & -8.2 & 13.2 & -0.8 \\ -5.7 & -3.7 & 6.2 & 7.5 \\ 0.1 & 0.0 & 0.1 & 0.4 \\ -0.1 & -0.0 & 0.1 & 0.3 \end{bmatrix},
$$

$$
\overline{D}_{qp} = 10^{-3} \begin{bmatrix} -30 & -4.8 & 22.5 & 15.9 & -8.1 & -0.1 & -0.0 \\ 12.0 & -18 & -30 & -22 & 13.7 & 0.6 & 0.6 \\ -27 & -15 & -10.5 & 1.1 & -1.3 & -0.5 & 0.9 \\ -4.8 & -3.4 & -2.3 & 3.1 & -0.5 & -0.3 & 0.2 \\ 2.2 & -7.2 & -14 & 8.5 & 2.6 & -1.1 & 0.2 \\ -0.0 & -0.0 & -0.0 & 0.0 & -0.0 & -0.0 & 0.0 \\ -0.1 & -0.1 & -0.0 & 0.1 & -0.1 & -0.1 & -0.0 \end{bmatrix}.
$$

Chapter 5. Stabilization of Linear Discrete-Time Systems with Saturating Controls and Norm-Bounded Time-Varying Uncertainty

Sophie Tarbouriech and Germain Garcia

L.A.A.S. - C.N.R.S.,
7 avenue du colonel Roche, 31077 Toulouse cedex 4, France.
E-mail : tarbour@laas.fr , garcia@laas.fr

1. Introduction

In practical control problems, numerous constraints have to be handled in order to design controllers which operate in a real environment. Concerning the system modelling step, usually the model is uncertain, the uncertainty resulting from simplifications or simply from parameter inaccuracies [6]... Concerning the control, in most of the pratical situations, the control is bounded and saturations can occur [4].

Usually, these two specificities are considered separately and a large amount of literature deals with them. For robust control, see [6], [3], [7] and references therein. For the saturating controls see for example [4], [5], [15] and references therein.

From a practical point of view, it would be interesting to elaborate some specific methods which take into account these two aspects and then which allow to design efficient controllers.

The aim of this chapter is to address this problem. Some preliminary results are proposed for the problem of robust stabilization of linear discrete-time systems subject to norm bounded uncertainty and bounded controls. The concepts of robust local, semi-global or still global stabilization are considered. To solve these, the quadratic stabilizability concept is used. In the local context, the idea is to derive both a stabilizing state feedback matrix K and a set of local stability as large as possible. In the semi-global context, that is, when the set of admissible initial states is an a priori given bounded set, some preliminary results are furnished. Finally, in the global context, that is, when the set of admissible initial states is the whole state space some remarks to show the possible restriction of the results are expressed.

The chapter is organized as follows. In the following section, the problem is stated, the principal notations and concepts introduced. Section 3 deals with the robust local stabilization with saturations. An algorithm based on simple calculations is proposed. In subsection 3.4, the connection between the results of subsection 3.2 and the disturbance rejection problem is investigated. The class of perturbations which can be rejected with a saturated control is characterized. Finally, these results are applied to a discretized model of an inverted pendulum. Sections 4 and 5 address the robust semi-global stabilization and robust global problem, respectively. Finally, section 6 presents some concluding remarks and perspectives.

2. Problem formulation

Consider the following discrete-time system :

$$x_{k+1} = (A + \Delta A)x_k + (B + \Delta B)u_k \tag{2.1}$$

where $x_k \in \Re^n$, $u_k \in \Re^m$, and matrices A, B are of appropriate dimensions. We suppose that system (2.1) satisfies the following assumptions :

(P1). The control vector u_k belongs to a compact set $\Omega \subset \Re^m$ defined by

$$\Omega = \{u \in \Re^m; -w \le u \le w\} \tag{2.2}$$

where w is a positive vector of $\Re^m$. By convention, the inequalities are componentwise.

(P2). Pair (A, B) is assumed to be stabilizable.

(P3). Matrices ΔA and ΔB are defined as

$$\begin{bmatrix} \Delta A & \Delta B \end{bmatrix} = DF(k) \begin{bmatrix} E_1 & E_2 \end{bmatrix} \tag{2.3}$$

where D, E_1 and E_2 are constant matrices of appropriate dimensions defining the structure of uncertainty. Matrix $F(k) \in \mathcal{F} \subset \Re^{l \times r}$ is the parameter uncertainty, where $\mathcal{F}$ is defined by

$$\mathcal{F} = \{F(k) \in \Re^{l \times r}; F(k)^t F(k) \le 1\} \tag{2.4}$$

This kind of uncertainty is the so-called "norm bounded uncertainty" and was considered frequently in the robust control literature : see [6], [7] and references therein.

By implementing a saturated state feedback $u_k = sat(Kx_k)$, $K \in \Re^{m \times n}$, system (2.1) becomes :

$$x_{k+1} = (A + \Delta A)x_k + (B + \Delta B)sat(Kx_k) \tag{2.5}$$

where each component of the saturation term

$$sat(Kx_k) = \begin{bmatrix} sat(Kx_k)^1 & \dots & sat(Kx_k)^m \end{bmatrix}^t$$

is defined by :

$$sat(Kx_k)^i = \begin{cases} w_i & \text{if} \quad K_i x_k > w_i \\ K_i x_k & \text{if} \quad -w_i \le K_i x_k \le w_i \\ -w_i & \text{if} \quad K_i x_k < -w_i \end{cases} , \quad \forall i = 1, ..., m \tag{2.6}$$

with K_i and w_i denotes respectively the ith line and the ith component of matrix K and vector w. Hence, system (2.5) is nonlinear.

If the controls do not saturate, that is, if $Kx_k \in \Omega$, or equivalently $x_k \in S(K, w)$ described by

$$S(K, w) = \{x_k \in \Re^n; -w \le Kx_k \le w\} \qquad (2.7)$$

system (2.5) admits the following linear model :

$$x_{k+1} = (A + \Delta A)x_k + (B + \Delta B)Kx_k \qquad (2.8)$$

The model of the closed-loop system is linear only inside $S(K, w)$. Nevertheless, it does not imply that any trajectory initiated in this set is a trajectory of the linear system (2.8).

If no particular hypothesis is done about matrix A, the problem addressed consists in studying the local stabilization of the saturated perturbed system (2.5) : local stabilization in the sense we have to determine a safe set $\mathcal{D}_0$ of initial conditions such that $\forall x_0 \in \mathcal{D}_0$ the resulting trajectory of system (2.5), denoted $x(k; x_0)$, asymptotically converges to the origin. When $\mathcal{D}_0$ is an a priori given (arbitrarily large) bounded set, the approach consists in finding a stabilizing state feedback matrix K such that $\mathcal{D}_0$ is a stability domain. It is referenced as the semi-global stabilization. But this approach requires some stability properties for the open-loop system [1], [5], [9]. Moreover, when $\mathcal{D}_0 = \Re^n$, the approach may be referred to the global stabilization, which requires some stability properties for the open-loop system [15].

Furthermore, it is well-known that for the stabilization of system (2.1) by linear state feedback, the concept of quadratic stabilizability [3] plays a central role [7]. In the current case, from the control constraints (2.2), the saturated state feedback (2.6) and without any stability hypothesis for matrix A, the study of the local stabilization of system (2.1) requires to investigate the quadratic stabilizability in a local way.

Hence, let us now define the concept of local quadratic stabilizability.

Definition 2.1. *System (2.1) is said to be locally quadratically stabilizable if there exists a static state feedback matrix $K \in \Re^{m \times n}$, a positive-definite symmetric matrix $P \in \Re^{n \times n}$ such that the following condition holds :*

$$\begin{aligned} L(x_k, k) = [(A + \Delta A)x_k + (B + \Delta B)sat(Kx_k)]^t P[(A \\ + \Delta A)x_k + (B + \Delta B)sat(Kx_k)] - x_k^t P x_k < 0 \end{aligned} \qquad (2.9)$$

for all pairs $(x_k, k) \in \mathcal{D}_0 \times \mathcal{N}$ and $F(k) \in \mathcal{F}$.

It is to be noted that this definition is not the same as the definition given in [3]. Here the decrease of the Lyapunov function has to be negative in a certain domain $\mathcal{D}_0$ of the state space. This restriction implies the notion of local quadratic stabilizability. If the domain $\mathcal{D}_0$ is equal to $\Re^n$ then the notion of global quadratic stabilization is considered.

Checking local quadratic stabilizability consists in looking for a quadratic Lyapunov function which is the same for all the systems in the uncertainty domain with respect to the control constraints.

Note also that in the case $x_k \in \mathcal{D}_0 \subseteq \mathcal{S}(K, w)$ condition (2.9) can be simply written as :

$$
\begin{aligned}
L(x_k, k) = [(A + \Delta A)x_k + (B + \Delta B)K x_k]^t P[(A + \Delta A)x_k \\
+ (B + \Delta B)K x_k] - x_k^t P x_k < 0
\end{aligned}
\tag{2.10}
$$

The key for the development of our results is the notion of invariant sets. A nonempty set $\mathcal{S} \subset \Re^n$ is positively invariant for system (2.5) if for any initial condition $x_0 \in \mathcal{S}$ then for all $k \geq 0$, $x_k \in \mathcal{S}$, for all admissible uncertainty $F(k) \in \mathcal{F}$. According to Definition 2.1, we can define the concept of positive invariant and contractive set used in this chapter.

Definition 2.2. *Let $V(x_k) = x_k^t P x_k$ be the quadratic Lyapunov function candidate for system (2.5). The set $\mathcal{D}_0 = \mathcal{S}(P, \gamma)$ defined by*

$$
\mathcal{S}(P, \gamma) = \{x \in \Re^n; x^t P x \leq \gamma; \gamma > 0\}
\tag{2.11}
$$

is positively invariant and contractive for system (2.5) if condition (2.9) holds $\forall x_k \in \mathcal{S}(P, \gamma)$ and $\forall F(k) \in \mathcal{F}$.

The objectives of this chapter are listed below.

A. In the local context.

1. Find a static state feedback matrix K and a domain $\mathcal{D}_0$ for which system (2.1) is locally quadratically stabilizable (in the sense of Definition 2.1).
2. Relate the local quadratic stabilizability conditions to an $\mathcal{H}_\infty$ norm constraint (disturbance rejection problem) as in the non-bounded control case.

B. In the semi-global context.
When the set $\mathcal{D}_0$ is an a priori given bounded set, some preliminary results are expressed.

C. In the global context.
When one wish to have $\mathcal{D}_0 = \Re^n$, some results on the global stabilization are expressed.

3. Robust local stabilization

3.1 Preliminaries

Let us first give some equivalent models for system (2.5). Thus, system (2.5) can be written

$$
\begin{aligned}
x_{k+1} &= (A + \Delta A)x_k + \sum_{i=1}^{m}\alpha^i(x_k)(B_i^c + \Delta B_i^c)K_i x_k \\
&= (\mathcal{A}(\alpha(x_k)) + \Delta\mathcal{A}(\alpha(x_k)))x_k
\end{aligned}
\tag{3.1}
$$

where B_i^c (ΔB_i^c) is the ith column vector of B (ΔB) and K_i is the ith row vector of K. The components $\alpha^i(x_k)$, $i = 1, ..., m$, of vector $\alpha(x_k)$ are defined by

$$
\alpha^i(x_k) = \begin{cases} \frac{w_i}{K_i x_k} & \text{if} \quad K_i x_k > w_i \\ 1 & \text{if} \quad -w_i \leq K_i x_k \leq w_i \\ -\frac{w_i}{K_i x_k} & \text{if} \quad K_i x_k < -w_i \end{cases}
\tag{3.2}
$$

and satisfy

$$
0 < \alpha^i(x_k) \leq 1
\tag{3.3}
$$

The control objective mainly concerns the determination of a domain of local stability. Then the domain of evolution of the state has to be limited and therefore the interval of evolution of the scalars $\alpha^i(x_k)$ has also to be limited.

If we compute a stabilizing state feedback K for system (2.1), it is not possible due to the time-varying terms and the saturation term to determine analytically the region of attraction of the origin [12]. Hence the determination of a positively invariant and contractive set for system (2.5) appears as an interesting way to approximate it.

Therefore, consider such a set $\mathcal{D}_0$. It follows that for any $x_k \in \mathcal{D}_0$, a lower bound for vector $\alpha(x_k)$ can be defined as follows :

$$
\alpha_0^i(x_k) = \min\{\alpha^i(x_k) \; ; \; \forall x_k \in \mathcal{D}_0\}, \; \forall i = 1, ..., m
\tag{3.4}
$$

Hence, in $\mathcal{D}_0$, the scalars $\alpha^i(x_k)$ defined in (3.2) satisfy :

$$
\alpha_0^i \leq \alpha^i(x_k) \leq 1 \; , \; \forall i = 1, ..., m
\tag{3.5}
$$

Note that when $\mathcal{D}_0 = \Re^n$, $\alpha^i(x_k)$ is allowed to vary as in (3.3) and therefore the global stabilization of system (2.5) is studied. In this case some stability properties for the open-loop system $(u_k = 0)$ are required.

From (3.4), we can represent the saturated system (2.5) by another model for each $x_k \in \mathcal{D}_0$. Based on the difference inclusions results [2], [11], for each $x_k \in \mathcal{D}_0$ system (2.5) can be represented by :

$$
x_{k+1} = \sum_{j=1}^{2^m} \lambda^j(x_k)(A^j(\alpha_0) + \Delta A^j(\alpha_0))x_k
\tag{3.6}
$$

with

$$
\sum_{j=1}^{2^m} \lambda^j(x_k) = 1, \; \lambda^j(x_k) \geq 0, \; \forall j.
\tag{3.7}
$$

The matrices $A^j(\alpha_0)$ and $\Delta A^j(\alpha_0)$ are defined as follows :

$$
\begin{aligned}
A^j(\alpha_0) &= A + B\Gamma(\beta_j)K & (3.8) \\
\Delta A^j(\alpha_0) &= \Delta A + \Delta B\Gamma(\beta_j)K & (3.9)
\end{aligned}
$$

for all $j = 1, ..., 2^m$. Matrix $\Gamma(\beta_j)$ is a positive diagonal matrix of $\Re^{m \times m}$ composed of the scalars β_{ji} which can take the value α_0^i or 1, for $i = 1, ..., m$ [5].

It is worth to notice that the vector α_0 defined in (3.4) allows to generate the polyhedral set $\mathcal{S}(K, \alpha_0) \subset \Re^n$ defined as :

$$
\mathcal{S}(K, \alpha_0) = \{x \in \Re^n; -\frac{w_i}{\alpha_0^i} \leq K_i x_k \leq \frac{w_i}{\alpha_0^i}; \alpha_0^i > 0, \forall i = 1, ..., m\} \quad (3.10)
$$

$\mathcal{S}(K, \alpha_0)$ contains $\mathcal{D}_0$ and is the maximal set in which for any $x_k \in \mathcal{S}(K, \alpha_0)$, the scalars $\alpha^i(x_k)$ defined in (3.2) satisfy (3.5). Hence, model (3.6) is only valid for states x_k belonging to $\mathcal{S}(K, \alpha_0)$. Clearly for $x_k \in \mathcal{S}(K, \alpha_0)$, matrix $\mathcal{A}(\alpha(x_k)) + \Delta\mathcal{A}(\alpha(x_k))$ in (3.1) belongs to a convex polyhedron of matrices whose vertices are $A^j(\alpha_0) + \Delta A^j(\alpha_0)$. The trajectories of system (2.5) or (3.1) are represented by those of system (3.6) so long as for some $x_0 \in \mathcal{S}(K, \alpha_0)$ one gets $x(k, x_0) \in \mathcal{S}(K, \alpha_0)$, $\forall k$, that is, if matrices $\mathcal{A}(\alpha(x(k, x_0))) + \Delta\mathcal{A}(\alpha(x(k; x_0)))$ in (3.1) belong to the convex hull $co\{A^1(\alpha_0) + \Delta A^1(\alpha_0), ..., A^\ell(\alpha_0) + \Delta A^\ell(\alpha_0)\}$, with $\ell = 2^m$. This corresponds to the fact that $\mathcal{S}(K, \alpha_0)$ must be a positively invariant set with respect to the trajectories of (3.6), which is not generally the case. Hence, the importance of the existence of a positively invariant set $\mathcal{D}_0 \subset \mathcal{S}(K, \alpha_0)$ appears in order for system (3.6) to represent system (2.5) or (3.1).

3.2 Robust local stabilization

A necessary and sufficient condition for the quadratic stabilizability of system (2.1) by linear state feedback when the control vector is not bounded is given in the following theorem [8].

Theorem 3.1. *System (2.1), without control constraints, is quadratically stabilizable by the static state feedback :*

$$
\begin{aligned}
K = &-(R_\epsilon + B^t(P^{-1} - \epsilon DD^t)^{-1}B)^{-1}B^t(P^{-1} - \epsilon DD^t)^{-1}A \\
&-\epsilon^{-1}(R_\epsilon + B^t(P^{-1} - \epsilon DD^t)^{-1}B)^{-1}E_2^t E_1
\end{aligned} \quad (3.11)
$$

if and only if there exist $\epsilon > 0$ and a positive definite matrix $P \in \Re^{n \times n}$ satisfying the following discrete Riccati equation

$$
\begin{aligned}
&\bar{A}^t(P^{-1} + BR_\epsilon^{-1}B^t - \epsilon DD^t)^{-1}\bar{A} - P \\
&+\epsilon^{-1}E_1^t(I_r - \epsilon^{-1}E_2 R_\epsilon^{-1} E_2^t)E_1 + Q = 0
\end{aligned} \quad (3.12)
$$

with

$$
\epsilon^{-1}I_l - D^t PD > 0 \quad (3.13)
$$

where $Q \in \Re^{n \times n}$ and $R \in \Re^{m \times m}$ are positive definite symmetric matrices and matrices $\bar{A}$ and R_ϵ are defined as :

$$R_\epsilon = R + \epsilon^{-1} E_2^t E_2 \qquad (3.14)$$
$$\bar{A} = A - \epsilon^{-1} B R_\epsilon^{-1} E_2^t E_1 \qquad (3.15)$$

Hence, by considering the case of system (2.1) with control constraints, to avoid the saturations on the controls, Theorem 3.1 can be used in order to determine a positively invariant and contractive set $\mathcal{D}_0$ included in $\mathcal{S}(K, w)$. It suffices, for example, to determine the largest ellipsoid $\mathcal{S}(P, \mu)$ defined by

$$\mathcal{S}(P, \mu) = \{x \in \Re^n; x^t P x \leq \mu; \mu > 0\} \qquad (3.16)$$

where the positive scalar μ is determined such that

$$\mathcal{S}(P, \mu) \subseteq \mathcal{S}(K, w) \qquad (3.17)$$

However, we are mainly interested by the possibility for the controls to saturate. Therefore we are interested in determining a positively invariant and contractive set in the sense of Definition 2.2.

Proposition 3.1. *Assume that there exist $\epsilon > 0$ and a positive definite matrix $P \in \Re^{n \times n}$ satisfying the discrete Riccati equation (3.12) and condition (3.13). System (2.1) is locally quadratically stabilizable by the static state feedback matrix K defined in (3.11) in the domain $\mathcal{S}(P, \gamma)$ if for all $j = 1, ..., 2^m$, one gets :*

$$\begin{aligned} -Q + K^t[(I_m - \Gamma(\beta_j))(R_\epsilon + B^t W B)(I_m - \Gamma(\beta_j)) \\ -\Gamma(\beta_j) R \Gamma(\beta_j)] K < 0 \end{aligned} \qquad (3.18)$$

where $W = (P^{-1} - \epsilon D D^t)^{-1}$.

Proof : Compute the decrease of the quadratic Lyapunov function $V(x_k) = x_k^t P x_k$ along the trajectories of system (2.5). It follows :

$$\begin{aligned} L(x_k, k) = x_k^t [\sum_{j=1}^{2^m} \lambda^j(x_k)(A^j(\alpha_0) + \Delta A^j(\alpha_0))^t P \sum_{j=1}^{2^m} \lambda^j(x_k) \\ (A^j(\alpha_0) + \Delta A^j(\alpha_0)) - P] x_k \end{aligned}$$

Then from the convexity of the function $V(x_k)$ it follows :

$$\begin{aligned} L(x_k, k) \leq \sum_{j=1}^{2^m} \lambda^j(x_k) x_k^t [(A^j(\alpha_0) + \Delta A^j(\alpha_0))^t P(A^j(\alpha_0) \\ + \Delta A^j(\alpha_0)) - P] x_k \end{aligned}$$

By setting $L_j(x_k, k) = x_k^t[(A^j(\alpha_0) + \Delta A^j(\alpha_0))^t P(A^j(\alpha_0) + \Delta A^j(\alpha_0)) - P]x_k$
if $L_j(x_k, k) < 0$ for $j = 1, ..., 2^m$, it follows $L(x_k, k) < 0$. By setting $W = (P^{-1} - \epsilon DD^t)^{-1}$ it follows :

$$L_j(x_k, k) \leq x_k^t[A^j(\alpha_0)^t W A^j(\alpha_0) - P \\ +\epsilon^{-1}(E_1 + E_2\Gamma(\beta_j)K)^t(E_1 + E_2\Gamma(\beta_j)K)]x_k$$

Then from (3.11), (3.12), (3.13) and (3.8), (3.9) one obtains :

$$L_j(x_k, k) \leq x_k^t[-Q + K^t[(\Gamma(\beta_j)(\epsilon^{-1}E_2^t E_2 + B^t W B)\Gamma(\beta_j) \\ +(R_\epsilon + B^t W B) - \Gamma(\beta_j)(R_\epsilon + B^t W B) \\ -(R_\epsilon + B^t W B)\Gamma(\beta_j)]K]x_k$$

Hence if condition (3.18) holds, it follows that $L_j(x_k, k) < 0$ for all $j = 1, ..., 2^m$ and therefore one gets $L(x_k, k) < 0$ for all $x_k \in \mathcal{S}(P, \gamma)$. This domain is positively invariant and contractive with respect to system (3.6) in the sense of Definition 2.2. Then, since in $\mathcal{S}(P, \gamma)$ system (3.6) represents system (2.5), the local asymptotic stability is guaranteed in this domain for system (2.5). $\square$

Note that by considering $x_k \in \mathcal{S}(P, \gamma)$, from (3.4), we can define the resulting α_0^i as

$$\alpha_0^i = \min(\frac{w_i}{\sqrt{\gamma}\sqrt{K_i P^{-1} K_i^t}}, 1), \ \forall i = 1, ..., m \qquad (3.19)$$

and therefore the vectors β_j, $j = 1, ..., 2^m$. In fact to apply Proposition 3.1 we have to determine the suitable positive scalar γ and thus vector α_0. The procedure can be stated as follows.

Algorithm 3.1.

- Step 1 : Choose any positive definite symmetric matrices $Q \in \Re^{n \times n}$ and $R \in \Re^{m \times m}$ and a certain starting value ϵ_0 for ϵ.
- Step 2 : Determine whether the modified Riccati equation (3.12) has a positive definite symmetric solution satisfying (3.13). If such a solution exists, compute the state feedback matrix K from (3.11) and go to Step 4. Otherwise, go to Step 3.
- Step 3 : Take $\epsilon = \epsilon_1 < \epsilon_0$. If ϵ is less than some computational accuracy ϵ_l : stop. Otherwise, go to Step 2.
- Step 4 : Compute, $\forall i = 1, ..., m$,

$$\max_x x^t P x = \eta_i \\ \text{subject to } K_i x \leq w_i$$

and select $\mu = \min_i \eta_i = \min_i \frac{w_i^2}{K_i P^{-1} K_i^t}$.

– Step 5 : Find the points x_i^* solution of

$$\max K_i x$$
$$\text{subject to } x^t P x = \mu$$

Then one obtains : $K_i x_i^* = \sqrt{\mu} \sqrt{K_i P^{-1} K_i^t}$. Thus compute $K_i x_i^* = \frac{w_i}{\theta_i}$, that is, $\theta_i = \frac{w_i}{\sqrt{\mu} \sqrt{K_i P^{-1} K_i^t}}$, $\forall i = 1, ..., m$. By definition $\theta_i \geq 1, \forall i = 1, ..., m$.

– Step 6 : Choose an increment $\Delta\rho$ and iterate the parameter ρ from 1. At each stage compute the 2^m possible combinations for vector β_j from $\beta_{ji} = 1$ or $\min(\frac{\theta_i}{\rho}, 1)$, $\forall i = 1, ..., m$. Thus construct the diagonal matrices $\Gamma(\beta_j)$ for $j = 1, ..., 2^m$. Test if condition (3.18) holds : if it is satisfied, ρ is a suitable value, otherwise stop.

– Step 7 : Among the previous suitable values of ρ, take $\rho_M = \max \rho$ and compute $\gamma = \rho_M^2 \mu$; then one obtains :

$$\alpha_0 = \left[\min(\frac{\theta_1}{\rho_M}, 1) \quad ... \quad \min(\frac{\theta_m}{\rho_M}, 1) \right]^t$$
$$S(P, \gamma) = \{x \in \Re^n; x^t P x \leq \gamma\} \subset S(K, \alpha_0) \tag{3.20}$$

The degree of conservativeness of the previous approach is mainly due to the representation chosen for the saturated system. All the trajectories of system (2.5) can be represented by those of system (3.6), but the converse is not true. Hence all the conditions obtained from representation (3.6) are only sufficient. Furthermore, the presented approach corresponds to find a common quadratic Lyapunov function for all the vertices of the polyhedron of matrices $co\{A^1(\alpha_0) + \Delta A^1(\alpha_0), ..., A^\ell(\alpha_0) + \Delta A^\ell(\alpha_0)\}$, $\ell = 2^m$, which is already restrictive.

A global optimization problem can be defined in order to determine simultaneously matrix P and scalar γ :

$$\min log(det(\gamma P^{-1}))$$

subject to
$$\begin{cases} P > 0 \\ \gamma > 0 \\ \alpha_0^i = \min(\dfrac{w_i}{\sqrt{\gamma}\sqrt{K_i P^{-1} K_i^t}}, 1), \ \forall i = 1, ..., m \\ \text{relation (3.11)} \\ \text{relation (3.12)} \\ \text{relation (3.13)} \\ \text{relation (3.18)} \end{cases}$$

Such an optimization problem is non-convex and therefore a simple resolution does not exist at the current time.

The existence of a positive definite symmetric matrix P solution of the Riccati equation (3.12) is independent of the choice of the positive definite matrices Q and R. This choice could be used in order to obtain additional requirements for the controlled system. This possibility is not investigated in the chapter.

3.3 Approach by Linear Matrices Inequalities (L.M.I.)

In order to propose some answers to problems above mentioned, an approach by L.M.I. can be proposed.

Let G_i be the ith row of an m-order identity matrix. Then the following proposition can be stated.

Proposition 3.2. *If there exist a symmetric positive definite matrix $S \in \Re^{n \times n} i$, a positive scalar ψ and a matrix $Y \in \Re^{m \times n}$ satisfying both :*

$$
\begin{bmatrix}
-S + DD^t & AS + B\Gamma(\beta_j)Y & 0 \\
(AS + B\Gamma(\beta_j)Y)^t & -S & (E_1 S + E_2\Gamma(\beta_j)Y)^t \\
0 & E_1 S + E_2\Gamma(\beta_j)Y & -I_r
\end{bmatrix} < 0
$$
$$\forall j = 1, ..., 2^m \tag{3.21}$$

$$
\begin{bmatrix}
S & Y^t G_i^t \\
G_i Y & \psi(\frac{w_i}{\alpha_0^i})^2
\end{bmatrix} \geq 0 , \ \forall i = 1, ..., m \tag{3.22}
$$

Then $K = YS^{-1}$ makes locally quadratically stabilizable system (2.1) in the domain :

$$\mathcal{S}(S^{-1}, \gamma) = \{x \in \Re^n; x^t S^{-1} x \leq \gamma\}, \ \gamma = \frac{1}{\psi} \tag{3.23}$$

Proof : The satisfaction of L.M.I. (3.22) means that the ellipsoid $\mathcal{S}(S^{-1}, \gamma)$, defined in (3.23), is included in $\mathcal{S}(K, \alpha_0)$. The satisfaction of L.M.I. (3.21) means that $\forall j = 1, ..., 2^m$:

$$
\begin{aligned}
x_k^t[(A + \Delta A + (B + \Delta B)\Gamma(\beta_j)YS^{-1})S(A + \Delta A \\
+ (B + \Delta B)\Gamma(\beta_j)YS^{-1})^t - S]x_k < 0
\end{aligned} \tag{3.24}
$$

for all $x_k \in \Re^n$, $x_k \neq 0$. Recall that model (3.6) is only valid in $\mathcal{S}(K, \alpha_0)$. Therefore the satisfaction of L.M.I. (3.21) and (3.22) implies that (3.24) is verified for all $x_k \in \mathcal{S}(S^{-1}, \gamma) \subset \mathcal{S}(K, \alpha_0)$, $x_k \neq 0$. Thus, the set $\mathcal{S}(S^{-1}, \gamma)$ is positively invariant and strictly contractive for the trajectories of system (3.6), and therefore for the trajectories of system (2.5). $\square$

The main difficulty regarding the application of Proposition 3.2 resides in the necessary a priori choice of the vector α_0, that is, of the different matrices $\Gamma(\beta_j)$. Indeed, one has to choose m positive scalars α_0^i, $0 < \alpha_0^i \leq 1$, $\forall i = 1, ..., m$, to compute the different matrices $\Gamma(\beta_j)$ and then to test if solutions S, ψ and Y, satisfying (3.21) and (3.22) exist. When such solutions do not exist, the difficulty consists in choosing among the m previous scalars α_0^i which are those to decrease and also how to decrease them.

Hence, a solution can be to consider an homogeneous saturation, that is, to consider that $\alpha_0^i = \zeta_0$, $\forall i = 1, ..., m$. In other words, the allowed saturation tolerance is the same in the m directions of the controls vector. It is clear that condition obtained will be more conservative but also less difficult numerically.

In such an approach via L.M.I., some supplementary conditions can be added as some given performance requirements when the closed-loop system does not saturate, that is, in $\mathcal{S}(K, w)$: for example, by placing the eigenvalues of $(A + \Delta A + (B + \Delta B)K)$ in a specific region of the complex plane. Furthermore, when the set $\mathcal{D}_0$ of initial conditions is an a priori given bounded set, the condition of the inclusion of $\mathcal{D}_0$ in $\mathcal{S}(S^{-1}, \gamma)$ has to be added in order to satisfy : $\mathcal{D}_0 \subset \mathcal{S}(S^{-1}, \gamma) \subset \mathcal{S}(K, \alpha_0)$.

3.4 Relation with the disturbance rejection problem

This section proposes an interpretation of the quadratic stabilizability condition in terms of $\mathcal{H}_\infty$ norm constraint (disturbance rejection problem).

Let us define the following system :

$$\begin{cases} x_{k+1} = Ax_k + Bu_k + Dw_k \\ z_k = E_1 x_k + E_2 u_k \end{cases} \tag{3.25}$$

where the control u_k satisfies (2.2)-(2.6). The vector $w_k \in L_2[0, \infty)$ is a perturbation and z_k is the controlled output.

We are concerned by designing a control $u_k = sat(Kx_k)$ such that the closed-loop system

$$\begin{cases} x_{k+1} = Ax_k + Bsat(Kx_k) + Dw_k \\ z_k = E_1 x_k + E_2 sat(Kx_k) \end{cases} \tag{3.26}$$

is locally stable and satisfies a given $\mathcal{H}_\infty$ norm constraint on disturbance attenuation. This can be formalized in the following definition.

Definition 3.1. *Let the positive constant η be given. Then, system (3.26) is said to be locally stabilizable with an $\mathcal{H}_\infty$ norm bound η if there exists a static state feedback matrix K such that :*

(i) the closed-loop system (3.26) is locally stabilizable in the domain $\mathcal{S}(P, \gamma)$,
(ii) subject to the assumption of the zero initial condition, the controlled output z_k satisfies

$$\| z_k \|_2 \leq \eta \| w_k \|_2$$

Without loss of generality, by introducing a scaling factor on matrices D, E_1 and E_2, we can take $\eta = 1$, that is, condition (ii) of Definition 3.1 becomes $\| z_k \|_2 \leq \| w_k \|_2$.

First, in order to satisfy the point (i) of Definition 3.1, we have to prove that, with respect to system (3.26), for any $x_k \in \mathcal{S}(P, \gamma)$ it follows that $x_{k+1} \in \mathcal{S}(P, \gamma)$. That means that vector w_k has also to belong to a compact set. Hence one needs the following lemma.

Lemma 3.1. *The domain $\mathcal{S}(P, \gamma)$ described in Algorithm 3.1 is positively invariant with respect to system (3.26) if*

$$w_k \in \mathcal{S}(D^t P D, \bar{\gamma}) = \{w_k \in \Re^l; w_k^t D^t P D w_k \leq \bar{\gamma}\}$$
$$\text{with } \bar{\gamma} \leq (1 - \max_{j=1,\dots,2^m} \lambda_{max}^{1/2}(A^j(\alpha_0)^t P A^j(\alpha_0)P^{-1}))^2 \gamma \qquad (3.27)$$

where $\lambda_{max}(.)$ denotes the maximal eigenvalue of the matrix.

Remark 3.1. Proposition 3.1 implies that $x_k^t(A^j(\alpha_0)^t P A^j(\alpha_0) - P)x_k < 0$ for all $x_k \in \mathcal{S}(P, \gamma)$ and therefore one gets $\lambda_{max}(A^j(\alpha_0)^t P A^j(\alpha_0)P^{-1}) < 1$.

Proof : We have to prove, relative to system (3.26), that for any $x_k \in \mathcal{S}(P, \gamma)$ it follows $x_{k+1} \in \mathcal{S}(P, \gamma)$. From Section 3 we can write system (3.26) as :

$$x_{k+1} = \sum_{j=1}^{2^m} \lambda^j(x_k) A^j(\alpha_0)x_k + Dw_k$$
$$z_k = \sum_{j=1}^{2^m} \mu^j(x_k) E^j(\alpha_0)x_k \qquad (3.28)$$

where the positive scalars μ^j are defined as λ^j in (3.7), matrix $A^j(\alpha_0)$ is defined in (3.8) and matrix $E^j(\alpha_0)$ is defined in the same way, that is,

$$E^j(\alpha_0) = E_1 + E_2 \Gamma(\beta_j)K \qquad (3.29)$$

Then, as in the proof of Proposition 3.1, considering the quadratic Lyapunov function $V(x_k) = x_k^t P x_k$ it follows :

$$\begin{aligned} V(x_{k+1}) \quad &\leq \quad \sum_{j=1}^{2^m} \lambda^j(x_k)(x_k^t A^j(\alpha_0)^t P A^j(\alpha_0)x_k \\ &\quad + 2w_k^t D^t P A^j(\alpha_0)x_k) + w_k^t D^t P D w_k \\ &= \sum_{j=1}^{2^m} \lambda^j(x_k)V_j(x_{k+1}) \end{aligned}$$

where $V_j(x_{k+1})$ is defined by

$$V_j(x_{k+1}) = x_k^t A^j(\alpha_0)^t P A^j(\alpha_0)x_k + 2x_k^t w_k^t D^t P A^j(\alpha_0)x_k + w_k^t D^t P D w_k$$

To prove the positive invariance of $\mathcal{S}(P, \gamma)$ with respect to system (3.28) we have to prove that for all x_k such that $V(x_k) \leq \gamma$ it follows $V(x_{k+1}) \leq \gamma$. It suffices to satisfy $V_j(x_{k+1}) \leq \gamma \; \forall j = 1, ..., 2^m$ and $\forall x_k \in \mathcal{S}(P, \gamma)$. Then by setting $\nu_j = \lambda_{max}(A^j(\alpha_0)^t P A^j(\alpha_0)P^{-1})$ it follows :

$$V_j(x_{k+1}) \leq \nu_j V(x_k) + 2\nu_j^{1/2}V(Dw_k)^{1/2}V(x_k)^{1/2} + V(Dw_k)$$
$$= (\nu_j^{1/2}V(x_k)^{1/2} + V(Dw_k)^{1/2})^2$$

Thus if condition (3.27) holds and since $\nu_j \leq \max\limits_j \nu_j$ it follows for all $x_k \in S(P,\gamma)$:

$$V_j(x_{k+1}) \leq \nu_j\gamma + 2\nu_j^{1/2}\gamma^{1/2}(1 - \nu_j^{1/2})\gamma^{1/2} + (1 - \nu_j^{1/2})^2\gamma = \gamma$$

Hence if condition (3.27) is satisfied one obtains $V_j(x_{k+1}) \leq \gamma$ for all $V(x_k) \leq \gamma$ and therefore $V(x_{k+1}) \leq \gamma$. The positive invariance of $S(P,\gamma)$ is proven. $\square$

The following proposition can then be stated.

Proposition 3.3. *If for a positive definite matrix Q there exist $\epsilon > 0$ and a positive definite symmetric matrix P solution of (3.12) and (3.13) and if condition (3.18) is satisfied then system (3.26) is locally quadratically stabilizable in domain $S(P,\gamma)$ with an $\mathcal{H}_\infty$ norm bound 1 by the static state feedback matrix K given in (3.11), provided that $w_k \in S(D^tPD,\bar\gamma)$.*

Proof : Condition (i) of Definition 3.1 applied to system (3.26) is satisfied by Proposition 3.1, provided that w_k satisfies Lemma 3.1. We have to prove that condition (ii) of Definition 3.1 holds for system (3.26). Remark first that if (3.12) and (3.13) are met, it follows for $\tilde{P} = \epsilon P$:

$$\bar{A}^t(\tilde{P}^{-1} - DD^t + B\frac{R_\epsilon^{-1}}{\epsilon}B^t)^{-1}\bar{A} - \tilde{P} + E_1^t(I_r - \epsilon^{-1}E_2R_\epsilon^{-1}E_2^t)E_1 \leq 0 \tag{3.30}$$

with

$$I_l - D^t\tilde{P}D > 0 \tag{3.31}$$

Assume now that $x_0 = 0$ and introduce :

$$J = \sum_{k=0}^{\infty}z_k^t z_k - w_k^t w_k \tag{3.32}$$

Hence, since $x_0 = 0$, for non-zero $w_k \in L_2(0,\infty)$ one gets :

$$J = \sum_{k=0}^{\infty}(z_k^t z_k - w_k^t w_k + x_{k+1}^t\tilde{P}x_{k+1} - x_k^t\tilde{P}x_k) - x_\infty^t\tilde{P}x_\infty$$

By setting $A(\alpha) = A + B\Gamma(\alpha)K$ and $E(\alpha) = E_1 + E_2\Gamma(\alpha)K$, where $\Gamma(\alpha)$ is a diagonal matrix in which every diagonal element is defined by (3.2) and satisfies (3.5), it follows :

$$J = \sum_{k=0}^{\infty}x_k^t E(\alpha)^t E(\alpha)x_k - w_k^t w_k + w_k^t D^t\tilde{P}Dw_k$$
$$+ 2x_k^t A(\alpha)^t\tilde{P}Dw_k + x_k^t(A(\alpha)^t\tilde{P}A(\alpha) - \tilde{P})x_k$$
$$- x_\infty^t\tilde{P}x_\infty$$

that is,

$$
\begin{aligned}
\mathcal{J} = & \sum_{k=0}^{\infty} x_k^t E(\alpha)^t E(\alpha) x_k + x_k^t A(\alpha)^t \tilde{P} D (I_l - D^t \tilde{P} D)^{-1} \\
& D^t \tilde{P} A(\alpha) x_k + x_k^t (A(\alpha)^t \tilde{P} A(\alpha) - \tilde{P}) x_k \\
& - \sum_{k=0}^{\infty} (w_k - (I_l - D^t \tilde{P} D)^{-1} D^t \tilde{P} A(\alpha) x_k)^t (I_l - \\
& D^t \tilde{P} D)(w_k - (I_l - D^t \tilde{P} D)^{-1} D^t \tilde{P} A(\alpha) x_k) \\
& - x_\infty^t \tilde{P} x_\infty
\end{aligned}
$$

Finally, by setting $\tilde{W} = (\tilde{P}^{-1} - DD^t)^{-1}$, $\tilde{M} = (I_l - D^t \tilde{P} D)$ and $\tilde{w}_k = (w_k - \tilde{M}^{-1} D^t \tilde{P} A(\alpha) x_k)$ it follows :

$$
\begin{aligned}
\mathcal{J} = & \sum_{k=0}^{\infty} x_k^t [E(\alpha)^t E(\alpha) + A(\alpha)^t \tilde{W} A(\alpha) - \tilde{P}] x_k \\
& - \sum_{k=0}^{\infty} \tilde{w}_k^t \tilde{M} \tilde{w}_k - x_\infty^t \tilde{P} x_\infty
\end{aligned}
$$

From Proposition 3.1, condition (3.30) and Lemma 3.1 one gets

$$
\mathcal{J} < 0, \ \forall x_k \in \mathcal{S}(P, \gamma), \ \forall w_k \in \mathcal{S}(D^t P D, \bar{\gamma})
$$

□

Remark 3.2. In the case of bounded controls, it cannot be possible to affirm that $\mathcal{J} < 0$ for any non-zero vector $w_k \in L_2(0, \infty)$ because for system (3.26) one has to verify that for any $x_k \in \mathcal{S}(P, \gamma)$ one has $x_{k+1} \in \mathcal{S}(P, \gamma)$. That means that vector w_k has also to belong to a compact set $\mathcal{S}(D^t P D, \bar{\gamma})$.

Remark 3.3. From (3.13), one can write : $w_k^t D^t P D w_k < \lambda_{max}(D^t P D) w_k^t w_k$. Hence, according to Lemma 3.1 if the condition

$$
w_k^t w_k \leq \frac{\bar{\gamma}}{\lambda_{max}(D^t P D)} \tag{3.33}
$$

is satisfied then the domain $\mathcal{S}(P, \gamma)$ is positively invariant with respect to system (3.26). In the case where the control is not constrained, γ is infinite and therefore $\bar{\gamma}$ is too. The unique constraint on w_k is then to belong to $L_2[0, \infty)$.

3.5 Numerical example

Consider the model of the inverted pendulum with norm-bounded time-varying uncertainties and bounded control. By considering the sampling period $T = 0.03s$, the discretized model is described by the following data :

$$A = \begin{bmatrix} 1 & 0.0002 & -0.0568 & 0 \\ 0 & 1.0166 & 0.0189 & 0.0301 \\ 0 & -0.0083 & 0.9421 & -0.0001 \\ 0 & 1.1076 & 1.2508 & 1.0114 \end{bmatrix} ; B = \begin{bmatrix} 0.0034 \\ 0.0373 \\ -0.1142 \\ 2.4661 \end{bmatrix}$$

$$D = \begin{bmatrix} 0.0001 \\ 0.0150 \\ -0.0020 \\ 1.0000 \end{bmatrix} ; E_1 = \begin{bmatrix} 0 & -0.6129 & -0.7284 & -0.0051 \end{bmatrix}$$

$$E_2 = 0.5000$$

The matrix A is unstable since its spectrum is

$$\sigma(A) = \{1; 1.1926; 0.8254; 0.9522\}$$

The control constraint is described by $w = 1$.

By Choosing $Q = I_4$ and $R = 100$, the resolution of the Riccati equation (3.12) with the constraint (3.13) gives the following matrix P and positive scalar ϵ :

$$P = 10^3 * \begin{bmatrix} 0.3376 & -0.3403 & -0.6892 & -0.0528 \\ -0.3403 & 2.5017 & 2.7719 & 0.3116 \\ -0.6892 & 2.7719 & 3.4661 & 0.3583 \\ -0.0528 & 0.3116 & 0.3583 & 0.0426 \end{bmatrix} ; \epsilon = 0.0078$$

The computation of the state feedback matrix gives

$$K = \begin{bmatrix} 0.3431 & -2.8315 & -3.0183 & -0.4279 \end{bmatrix}$$

By applying Algorithm 3.1, it follows :

$$\mu = 193.7408 ; \theta = 1 ; \alpha_0 = 0.6546 ; \rho_M = 1.5277$$

One obtains $\gamma = 452.1497$. Then the domain $\mathcal{S}(P, \gamma)$ satisfies Definition 2.2. The choice of matrices Q and R has an influence on the obtained results. The role of these matrices has to be investigated in more details and this is an important open problem.

4. Robust semi-global stabilization

4.1 Main results

Consider now that the set of initial admissible states, denoted $\mathcal{D}_0$, is an a priori given bounded set. In the case without uncertainty (that is, when $\Delta A = 0$ and $\Delta B = 0$), it is now well-known that the semi-global stabilization can be obtained if the eigenvalues of A are located in the unit disk, that is, if matrix A is not strictly unstable [9], [1].

From Definition 2.1, one can define the notion of semi-globally quadratically stabilizable system as follows.

Definition 4.1. *System (2.1) is said to be semi-globally quadratically stabilizable in $\mathcal{D}_0$, a priori given, if there exists a static state feedback matrix $K \in \Re^{m \times n}$, a positive-definite symmetric matrix $P \in \Re^{n \times n}$ such that the following condition holds :*

$$L(x(k, x_0)) < 0 \ , \ \forall x_0 \in \mathcal{D}_0 \ , \ \forall k \in \mathcal{N} \ , \ F(k) \in \mathcal{F} \tag{4.1}$$

where $L(x(k; x_0))$ denotes the decrease of the quadratic function $x^t P x$ along the trajectories of system (2.5) initiating in $\mathcal{D}_0$.

Definition 4.1 means that $\mathcal{D}_0$ is a domain of stability for system (2.5), whereas in Definition 2.1 the considered domain $\mathcal{D}_0$ (and therefore the domain $\mathcal{S}(P, \gamma)$ determined in section 3) is a positively invariant and contractive set for system (2.5).

Suppose without loss of generality that the domain $\mathcal{D}_0$ a priori given is defined by :

$$\mathcal{D}_0 = \mathcal{S}(R_0, \rho_0) = \{x_0 \in \Re^n; x_0^t R_0 x_0 \le \rho_0\} \tag{4.2}$$

where $\rho_0 > 0$ and $R_0 \in \Re^{n \times n}$ is a symmetric positive definite matrix. Based on the description of system (2.5) described in subsection 3.1, the following proposition can be stated.

Proposition 4.1. *Assume that there exist $\epsilon > 0$ and a positive definite matrix $P \in \Re^{n \times n}$ satisfying the discrete Riccati equation (3.12) and condition (3.13). System (2.1) is semi-globally quadratically stabilizable by the static state feedback matrix K defined in (3.11) in the domain $\mathcal{S}(R_0, \rho_0)$ if for any positive scalar γ satisfying :*

$$\gamma > \rho_0 \lambda_{max}(R_0^{-1/2} P R_0^{-1/2}) \tag{4.3}$$

one gets for all $j = 1, ..., 2^m$:

$$-Q + K^t[(I_m - \Gamma(\beta_j))(R_\epsilon + B^t W B)(I_m - \Gamma(\beta_j)) - \Gamma(\beta_j) R \Gamma(\beta_j)]K < 0 \tag{4.4}$$

where $W = (P^{-1} - \epsilon D D^t)^{-1}$ and the 2^m possible matrices $\Gamma(\beta_j)$ are computed from the chosen scalar γ from (3.19).

Proof : If the scalar γ satisfies relation (4.3) then it follows that the matrix $R_0/\rho_0 - P/\gamma$ is positive definite ; hence, from Lemma 4.1 in [14], one gets $\mathcal{S}(R_0, \rho_0) \subset \mathcal{S}(P, \gamma)$. Furthermore, from the chosen scalar γ satisfying (4.3) one can compute the suitable matrices $\Gamma(\beta_j)$ from (3.19). Moreover, if condition (4.4) holds then from the proof of Proposition 3.1 it follows that $L_j(x_k, k) < 0$ for all $j = 1, ..., 2^m$ and therefore one gets $L(x_k, k) < 0$ for all $x_k \in \mathcal{S}(P, \gamma)$. Therefore, since $\mathcal{S}(R_0, \rho_0) \subset \mathcal{S}(P, \gamma)$ one gets $L(x_k, k) < 0$, $\forall x_0 \in \mathcal{S}(R_0, \rho_0)$ and $\forall F(k) \in \mathcal{F}$. $\square$

Remark 4.1. Suppose that the set of admissible initial states is a polyhedron defined by its vertices as :

$$\mathcal{D}_0 = co\{v_1, ..., v_s\} \tag{4.5}$$

Then Proposition 4.1 applied by choosing the positive scalar γ such that

$$\gamma < \min_{i=1,...,s} \left(\frac{1}{v_i^t P^{-1} v_i}\right) \tag{4.6}$$

4.2 Approach by L.M.I.

As in subsection 3.3, let G_i be the ith row of an m-order identity matrix. Then when the set of admissible initial states $\mathcal{D}_0$ is definied in (4.2), the following proposition can be stated.

Proposition 4.2. *If there exist a symmetric positive definite matrix $S \in \Re^{n \times n}$, a positive scalar ψ and a matrix $Y \in \Re^{m \times n}$ satisfying both :*

$$\begin{bmatrix} -S + DD^t & AS + B\Gamma(\beta_j)Y & 0 \\ (AS + B\Gamma(\beta_j)Y)^t & -S & (E_1 S + E_2\Gamma(\beta_j)Y)^t \\ 0 & E_1 S + E_2\Gamma(\beta_j)Y & -I_r \end{bmatrix} < 0$$
$$\forall j = 1, ..., 2^m \tag{4.7}$$

$$\begin{bmatrix} \psi R_0/\rho_0 & \psi I_n \\ \psi I_n & S \end{bmatrix} > 0 \tag{4.8}$$

$$\begin{bmatrix} S & Y^t G_i^t \\ G_i Y & \psi(\frac{w_i}{\alpha_0^i})^2 \end{bmatrix} \geq 0 , \ \forall i = 1, ..., m \tag{4.9}$$

Then $K = YS^{-1}$ makes semi-globally quadratically stabilizable system (2.1) in the domain $\mathcal{S}(R_0, \rho_0)$ such that :

$$\mathcal{S}(R_0, \rho_0) \subset \mathcal{S}(S^{-1}, \gamma) = \{x \in \Re^n; x^t S^{-1} x \leq \gamma\}, \ \gamma = \frac{1}{\psi} \tag{4.10}$$

Proof : The proof mimics the one of Proposition 3.2. Note that the satisfaction of L.M.I. (4.8) means that $\mathcal{S}(R_0, \rho_0) \subset \mathcal{S}(S^{-1}, \gamma)$. $\square$

When the set $\mathcal{D}_0$ is a polyhedral set described by its vertices as in (4.5), Proposition 4.2 applies by considering the following L.M.I. :

$$\begin{bmatrix} \psi & \psi v_i^t \\ \psi v_i & S \end{bmatrix} > 0 \ , \ \forall i = 1, ..., s \tag{4.11}$$

instead of L.M.I. (4.8). The satisfaction of L.M.I. (4.11) implies that $v_i^t S^{-1} v_i \leq \gamma$, with $\gamma = \frac{1}{\psi}$

4.3 Numerical example

Consider the crane system borrowed from [16]. By considering the sampling period $T = 0.01s$, the discretized system (2.1) is described by the following data :

$$A = \begin{bmatrix} 1 & 0.0098 & 0 & 0 \\ 0 & 0.9625 & 0 & 0 \\ 0 & -0.0001 & 0.9982 & 0.0100 \\ 0 & -0.0188 & -0.3606 & 0.9982 \end{bmatrix} ; B = \begin{bmatrix} 0.0006 \\ 0.1102 \\ 0.0003 \\ 0.0551 \end{bmatrix}$$

The control takes values in the set Ω defined in (2.2) by $w = 2$.
Consider matrices ΔA and ΔB defined in (2.3) described by :

$$D = \begin{bmatrix} -0.3 \\ 1 \\ -0.04 \\ 0.12 \end{bmatrix} ; E_1 = \begin{bmatrix} 0 & 0.021 & 0 & 0 \end{bmatrix} ; E_2 = 0.2$$

Assume that the set of initial admissible states is defined in (4.3) by $R_0 = I_4$ and $\rho_0 = 300$.
By choosing $Q = 10^{-3}I_4$ and $R = 1000$, the application of Proposition 4.1 gives

$$P = 10^3 * \begin{bmatrix} 0.1656 & 0.0372 & 0.0002 & 0 \\ 0.0372 & 0.0658 & 0.0443 & -0.0053 \\ 0.0002 & 0.0443 & 1.1111 & 0 \\ 0 & -0.0053 & 0 & 0.0308 \end{bmatrix} , \ \epsilon = 6.1035 \ 10^{-5}$$

$$K = \begin{bmatrix} -0.0025 & -0.0456 & -0.0029 & -0.0007 \end{bmatrix}$$
$$\gamma = 3.4005 \ 10^5$$

Hence one gets $\rho_0 * \lambda_{max}(P) = 3.3390 \ 10^5 < \gamma$.

5. Robust global stabilization

In the case without uncertainty (that is, when $\Delta A = 0$ and $\Delta B = 0$), it is now well-known that the global stabilization by a static linear state feedback can be obtained if the eigenvalues of A are stable [15], [10].

In Definition 2.1, the notion of globally quadratically stabilizability is obtained by considering that condition (2.9) holds $\forall x_k \in \Re^n$, $x_k \neq 0$, $\forall k \in \mathcal{N}$ and $F(k) \in \mathcal{F}$.

The following proposition can then be stated.

Proposition 5.1. *If there exist a symmetric positive definite matrix $P \in \Re^{n \times n}$ and a positive scalar ϵ solutions of :*

$$A^t(P^{-1} - \epsilon DD^t)^{-1}A - P + \epsilon^{-1}E_1^t E_1 + Q = 0 \qquad (5.1)$$

with

$$\epsilon^{-1}I_l - D^t PD > 0 \qquad (5.2)$$

where $Q \in \Re^{n \times n}$ is any symmetric positive definite matrix, then system (2.1) is globally quadratically stabilizable by the state feedback

$$K = -\Gamma(h)[B^t(P^{-1} - \epsilon DD^t)^{-1}A + \epsilon^{-1}E_2^t E_1] \qquad (5.3)$$

with the positive diagonal matrix $\Gamma(h) \in \Re^{m \times m}$ satisfies :

$$2 > \Gamma(h)^{\frac{1}{2}}[B^t(P^{-1} - \epsilon DD^t)^{-1}B + \epsilon^{-1}E_2^t E_2]\Gamma(h)^{\frac{1}{2}} > 0. \qquad (5.4)$$

Proof : Compute the decrease of the quadratic function $V(x_k) = x_k^t P x_k$ along the trajectories of system (2.5). By setting $F = F(k)$, $A(\alpha) = A + B\Gamma(\alpha)K$ and $E(\alpha) = E_1 + E_2\Gamma(\alpha)K$, where $\Gamma(\alpha)$ is a diagonal matrix whose the diagonal elements are defined in (3.2)-(3.3), one gets :

$$
\begin{aligned}
\Delta V_\alpha(x_k) &= x_k^t(A(\alpha) + DFE(\alpha))^t P(A(\alpha) + DFE(\alpha))x_k - x_k^t P x_k \\
&= x_k^t(A(\alpha)^t PA(\alpha) - P)x_k + x_k[A(\alpha)^t PDFE(\alpha) \\
&\quad + E(\alpha)^t F^t D^t PA(\alpha) + E(\alpha)^t F^t D^t PDFE(\alpha)]x_k
\end{aligned}
$$

From Lemma 5 in [7], if $\epsilon^{-1}I_l - D^t PD > 0$, $\epsilon > 0$ (condition (5.2) holds) then it follows :

$$
\begin{aligned}
\Delta V_\alpha(x_k) \leq\ & x_k^t[A(\alpha)^t PA(\alpha) - P + \epsilon^{-1}E(\alpha)^t E(\alpha) \\
&+ A(\alpha)^t PD(\epsilon^{-1}I_l - D^t PD)^{-1}D^t PA(\alpha)]x_k
\end{aligned}
$$

that is,

$$\Delta V_\alpha(x_k) \leq x_k^t[A(\alpha)^t(P^{-1} - \epsilon DD^t)^{-1}A(\alpha) - P + \epsilon^{-1}E(\alpha)^t E(\alpha)]x_k$$

From equation (5.1) one gets

$$P = A^t (P^{-1} - \epsilon DD^t)^{-1} A + \epsilon^{-1} E_1^t E_1 + Q$$

Then we replace $A(\alpha)$ by $A - B\Gamma(\alpha)\Gamma(h)[B^t(P^{-1} - \epsilon DD^t)^{-1}A + \epsilon^{-1}E_2^t E_1]$ from (5.3) we replace $B^t(P^{-1} - \epsilon DD^t)^{-1}A$ by $-\Gamma(h)^{-1}K - \epsilon^{-1}E_2^t E_1$. It follows :

$$\begin{aligned}\Delta V_\alpha(x_k) \leq\ & x_k^t K^t[-2\Gamma(\alpha)\Gamma(h)^{-1} + \Gamma(\alpha)(B^t(P^{-1} - \epsilon DD^t)^{-1}B \\ & + \epsilon^{-1} E_2^t E_2)\Gamma(\alpha)]Kx_k - x_k^t Q x_k\end{aligned}$$

It is easy to prove that $x_k^t K^t[-2\Gamma(\alpha)\Gamma(h)^{-1} + \Gamma(\alpha)(B^t(P^{-1} - \epsilon DD^t)^{-1}B + \epsilon^{-1}E_2^t E_2)\Gamma(\alpha)]Kx_k \leq 0$ from the condition (5.4), by noting that matrix $\Gamma(\alpha)\Gamma(h)^{-1}$ is positive definite. Therefore $\Delta V_\alpha(x_k) < 0$, $\forall x_k \in \Re^n$, $x_k \neq 0$.
□

Remark 5.1. The existence of a symmetric positive definite matrix P solution of (5.1), for any chosen symmetric and positive definite matrix Q, means that the matrix $A + \Delta A$ is asymptotically stable.

Remark 5.2. Along the trajectories of system (2.5), the decrease of the quadratic Lyapunov function $V(x_k) = x_k^t P x_k$ satisfies the following :

- $\Delta V_\alpha(x_k) < 0$, $\forall x_k \in \Re^n \backslash \{0\}$ when Q is positive definite.
- $\Delta V_\alpha(x_k) < 0$, $\forall x_k \in \Re^n \backslash \{0\}$ when Q is semi-definite positive with $(Ker(Q) \cap Ker(K)) \backslash \{0\} = \emptyset$.
- When Q is semi-definite positive with $(Ker(Q) \cap Ker(K)) \backslash \{0\} \neq \emptyset$, one needs to prove that this subspace is not invariant with respect to the motion of the open-loop system $x_{k+1} = (A + \Delta A)x_k$. The procedure is completely described in [13]. In fact one can show that to obtain the global stability of system (2.5), the matrices of uncertainty D, E_1 and E_2 must have a certain structure.

Remark 5.3. Due to the structure imposed on the state feedback matrix K (in (5.3)), the resolution via a L.M.I. approach seems to lose its interest. Nevertheless, it is clear that a first step may consist in computing a positive definite symmetric matrix $S \in \Re^{n \times n}$ solution of

$$\begin{bmatrix} -S + DD^t & AS & 0 \\ SA^t & -S & SE_1^t \\ 0 & E_1 S & -I_r \end{bmatrix} < 0 \tag{5.5}$$

and next to compute a state feedback matrix K as

$$K = -\Gamma(h)[B^t(S - DD^t)^{-1}A + E_1^t E_2] \tag{5.6}$$

where diagonal matrix $\Gamma(h) \in \Re^{m \times m}$ satisfies :

$$2 > \Gamma(h)^{\frac{1}{2}}[B^t(S - DD^t)^{-1}B + E_2^t E_2]\Gamma(h)^{\frac{1}{2}} > 0. \tag{5.7}$$

5.1 Numerical example

Consider the illustrative example borrowed from [5]. System (2.1) is described by the following data :

$$A = \begin{bmatrix} 1 & 0 & 0 \\ 0 & 0.9 & -0.06 \\ 0 & 0.06 & 0.9 \end{bmatrix} \; ; \; B = \begin{bmatrix} 1 & 0.4 \\ 2 & 0 \\ 3 & -1 \end{bmatrix}$$

The control takes values in the set Ω defined in (2.2) by $w = \begin{bmatrix} 1 \\ 2 \end{bmatrix}$.

Consider matrices ΔA and ΔB defined in (2.3) described by :

$$D = \begin{bmatrix} 0 \\ 1 \\ -1 \end{bmatrix} \; ; \; E_1 = \begin{bmatrix} 0 & 0.02 & -0.1 \end{bmatrix} \; ; \; E_2 = \begin{bmatrix} 1 & -1 \end{bmatrix}$$

By choosing $Q = \begin{bmatrix} 0 & 0 & 0 \\ 0 & 1 & 0 \\ 0 & 0 & 1 \end{bmatrix}$, the application of Proposition 5.1 gives :

$$P = \begin{bmatrix} 1 & 0 & 0 \\ 0 & 9.2432 & 4.2667 \\ 0 & 4.2667 & 41.4505 \end{bmatrix} \; ; \; \epsilon = 0.0031$$

$$K = \begin{bmatrix} -0.0015 & -0.0626 & -0.1471 \\ -0.0007 & 0.0211 & 0.0163 \end{bmatrix}$$

6. Conclusion

In this chapter, the stabilization of a linear discrete-time norm bounded uncertain system was addressed through a saturated control. The notions of robust local, semi-global and global stabilization were considered.

However, we have focused our study on the robust local stabilization problem. The concept of quadratic local stabilizability was defined and used to derive an algorithm allowing both the control law design, based on the existence of a positive definite symmetric matrix solution of a discrete Riccati equation, and a set of safe initial conditions for which asymptotic stability is guaranteed. The size of this set depends on the Riccati equation solution in a complex way and the problem of selecting a matrix which maximizes its size is not obvious and remains open. In this sense, the approach via the satisfaction of some L.M.I. conditions may be considered as an interesting way of research. Some connections with the disturbance rejection problem are also pointed out. The case of state feedbak was considered, but in practice, is often unrealistic. Output feedback stabilization is then more adequate. The results presented in this chapter are only preliminary, some problems have to be investigated in more details. Among them, the output feedback stabilization and the role of the Riccati equation solution on the size of the set of safe initial conditions will be addressed in a forthcomming issue.

References

1. J. Alvarez-Ramirez, R. Suarez, J. alvarez : *Semi-global stabilization of multi-input linear systems with saturated linear state feedback*, Systems & Control Letters, 23, pp.247-254, 1994.
2. J.P. Aubin and A. Cellina : *Differential inclusions*, Springe-Verlag, 1984.
3. B.R. Barmish : *Necessary and sufficient conditions for quadratic stabilizability of an uncertain system*, J. Optim. Theory Appl., vol.46, no.4, pp. , 1985.
4. D.S. Bernstein and A.N. Michel : *A chronological bibliography on saturating actuators*, Int. J. of Robust and Nonlinear Control, vol.5, pp.375-380, 1995.
5. C. Burgat and S. Tarbouriech : *Non-Linear Systems*, vol.2, Chapter 4, Annexes C, D , E, Chapman & Hall, London (U.K), 1996.
6. P. Dorato : *Robust Control*, IEEE Press Book, 1987.
7. G. Garcia, J. Bernussou, D. Arzelier : *Robust stabilization of discrete-time linear systems with norm-bounded time varying uncertainty*, Systems and Control Letters, vol.22, pp.327-339, 1994.
8. G. Garcia, J. Bernussou, D. Arzelier : *Disk pole location control for uncertain system with $\mathcal{H}_2$ guaranteed cost*, LAAS Report no.94216, submitted for review.
9. Z. Lin and A. Saberi : *Semi-global exponential stabilization of linear discrete-time systems subject to input saturation via linear feedbacks*, Systems and Control Letters, 24, pp.125-132, 1995.
10. C.C.H. Ma : *Unstability of linear unstable systems with inputs limits*, J. of Dynamic Syst., Measurement and Control, 113, pp.742-744, 1991.
11. A.P. Molchanov and E.S. Pyatniskii : *Criteria of asymptotic stability of differential and difference inclusions encountered in control theory*, Systems and Control Letters, 13, pp.59-64, 1989.
12. R. Suarez, J. Alvarez-Ramirez, J. Alvarez : *Linear systems with single saturated input : stability analysis*, Proc. of 30th IEEE-CDC, Brighton, England, pp.223-228, December 1991.
13. S. Tarbouriech and G. Garcia : *Global stabilization for linear discrete-time systems with saturating controls and norm-bounded time-varying uncertainty*, LAAS Report no.95112, submitted for review.
14. G.F. Wredenhagen and P.R. Bélanger : *Piecewise-linear LQ control for systems with input constraints*, Automatica, vol.30, no.3, pp.403-416, 1994.
15. Y. Yang : *Global stabilization of linear systems with bounded feedback*, Ph.D. Dissertation, New Brunswick Rutgers, the State University of New Jersey, 1993.
16. K. Yoshida and H. Kawabe : *A design of saturating control with a guaranteed cost and its application to the crane control system*, IEEE Trans. Autom. Control, vol.37, no.1, pp.121-127, 1992.

Chapter 6. Nonlinear Controllers for the Constrained Stabilization of Uncertain Dynamic Systems

Franco Blanchini[1] and Stefano Miani[2]

[1] Dipartimento di Matematica e Informatica, Università degli Studi di Udine, via delle Scienze 208, 33100 Udine - ITALY
[2] Dipartimento di Elettronica e Informatica, Università degli Studi di Padova, via Gradenigo 6/a, 35131 Padova - ITALY

1. Introduction

In the practical implementation of state feedback controllers there are normally several aspects which the designer has to keep in consideration and which impose restrictions on the allowable closed loop behavior. For instance a certain robustness of the closed loop system is desirable if not necessary to guarantee a stable functioning under different operating conditions which might be for example caused by effectively different set points, component obsolescence, neglected nonlinearities or high frequencies modes. Another issue which has surely to be taken into account is most often the presence of constraints on the control values and on the state variables. The former usually derives from saturation effects of the actuators whereas the latter normally comes from the necessity of keeping the states in a region in which the linearized model represents a good approximation of the real plant or might even be imposed by safety considerations.

The constrained control stabilization is by itself a challenging matter and in this contest the designer can either analyze the effects of saturating a stabilizing control law or he can include the constraints in the controller requirements. If stability is the only matter of concern then the first approach is indeed the easiest. The counterpart of this immediateness is unfortunately given by the extremely restricted set of initial states which can be asymptotically driven to the origin [20, 17], say the attraction set. Moreover if state constraints and uncertainties have to be considered then the first approach shows up its deficiencies so that the second approach appears definitely as the most preferable one. In this second class there are several techniques which can be followed to purse the desired performance specification while satisfying the imposed constraints and among these one of the approaches which can be used to overcome these limitations is that based on invariant regions [16, 15, 21, 25, 4, 22, 23, 1].

The key idea which lies behind this approach is that of determining a set of initial conditions starting from which the state evolution can be brought to the origin while assuring that no control and state constraint violation occur.

This is quite a standard approach and practically amounts to determine a candidate Lyapunov function for the constrained system which can be made decreasing along the system trajectories by a proper choice of the feedback control. Of course there is a certain freedom in the choice of such Lyapunov functions. From the existing literature it turns out that the class of quadratic functions has been the most investigated one mainly due to the elegant and powerful results existing in this area. Although this class is well established and capable of furnishing simple linear control laws, it is not perfectly suited for constrained control synthesis problem due to its conservativity. It is in fact possible to furnish examples of uncertain systems for which quadratic stabilizability cannot be provided although the system is stabilizable [6]. Furthermore it is known that the largest domain of attraction to the origin for a constrained control problem can be arbitrarily closely approximated (or even exactly determined) by means of polyhedrons while ellipsoidal sets may only provide rough approximations.

For these reasons in the last years several authors [12, 25, 23, 4, 22, 15] have put their attention on the class of polyhedral functions (say functions whose level surfaces are polyhedrons in $\mathbb{R}^n$) and the associated polyhedral invariant sets. These functions have their force in their capability of well representing linear constraints on state and control variables while being representable by a finite number of parameters. Moreover it has been recently shown [6] that this class is universal for the robust constrained stabilization problem if memoryless structured uncertainties are considered in the sense that in this case there exists a state feedback stabilizing control law if and only if there exists a polyhedral function which can be made decreasing along the system trajectories by a proper choice of the control law.

In this chapter we focus on the problem of determining a state feedback stabilizing control law for constrained dynamic systems (both in the continuous and discrete-time case) affected by structured memoryless uncertainties. In section 2. we will report some preliminary definitions and in the following section we will introduce the class of systems under consideration and we will state the problem. We will reformulate the problem in terms of Lyapunov functions in section 4. where we will also report some known results concerning the constrained stabilization of dynamic systems by means of polyhedral Lyapunov functions. We will present the necessary and sufficient conditions for the existence of a Lyapunov function which will be used in the sections 5. 6. to give a solution to the constrained stabilization problem. Then, based on these results, in section 7. we will focus our attention on the determination of a stabilizing feedback control law for the continuous and discrete-time case. In section 8. we will present an application of the proposed techniques to a two dimensional laboratory system and finally in section 9. we will report some final considerations and the directions for further research in this area.

Schematically, the outline of the present paper will be the following:

– Definitions

2. Definitions

We denote with $conv(S)$ the convex hull of a set $S \subseteq \mathbb{R}^n$. We will call C-set a closed and convex set containing the origin as an interior point. Given a C-set $P \subseteq \mathbb{R}^n$ we denote with $\lambda P = \{y = \lambda x, x \in P\}$ and with ∂P its border. Given r points in $\mathbb{R}^n$ we denote with $conv(y_1, .., y_r)$ their convex combination and given two C-sets P_1 and P_2 we denote by $conv(P_1, P_2) = conv(P_1 \bigcup P_2)$ their convex hull. We will be mostly dealing with polyhedral C-sets in view of their advantage of being representable by a finite set of linear inequalities. A polyhedral C-set can indeed be represented in terms of its delimiting planes as

$$P = \{x \in \mathbb{R}^n : F_i x \leq 1, i = 1, \ldots, s\},$$

where each F_i represents an n-dimensional row vector as well as by its dual representation

$$P = conv(v_1, .., v_k) = conv(V),$$

in terms of its vertex set $V = \{v_1\ v_2 \ldots v_r\}$, (or, with obvious meaning of the notation, its vertex matrix $V = [v_1\ v_2 \ldots v_r]$) which will be denoted by $vert\{P\}$. For these sets it is possible to introduce a compact notation using component-wise vector inequalities with which the set expression becomes

$$P = \{x : Fx \leq \bar{1}\}$$

or, using the dual notation,

$$P = \{x = V\alpha,\ \alpha \in \mathbb{R}^r,\ \|\alpha\|_1 \leq 1\},$$

where F is an $s \times n$ full column rank matrix having rows F_i and V is the full row rank matrix having the vertices v_i as column defined above, and $\bar{1} = [1 \ldots 1]^T$ represents an s-dimensional column vector. We say that a function $\Psi(x)$ from $\mathbb{R}^n$ to $\mathbb{R}^+$ is a Gauge function if the following properties hold:

$$
\begin{aligned}
\Psi(x) &> 0 & &\text{for every } x \neq 0, \\
\Psi(\lambda x) &= \lambda\Psi(x) & &\text{for every } \lambda \geq 0, \\
\Psi(x + y) &\leq \Psi(x) + \Psi(y) & &\text{for every } x,\ y \in \mathbb{R}^n.
\end{aligned}
\tag{2.1}
$$

Every C-set P naturally induces a Gauge function, the Minkowski functional $\Psi_P(x)$, whose expression is given by

$$\Psi_P(x) = \min\{\lambda \;:\; x \in \lambda P\}.$$

Accordingly to the above every C-set P can be seen as the unit ball of a proper Gauge function

$$P = \{x : \Psi_P(x) \le 1\}$$

Moreover a symmetric C-set P induces a 0-symmetric Gauge function $\Psi_P(x)$ (i.e. such that $\Psi(x) = \Psi(-x)$) which is a norm and every norm induces a symmetric C-set. For a polyhedral set $P = \{x : Fx \le \bar{1}\}$ the above expression can be simplified as

$$\Psi_P(x) = \max_i F_i x$$

and in this case we will denote by $I(x)$ the set of indexes for which $F_i x$ is maximum

$$I(x) = \{i : \; F_i x = \Psi_P(x)\}. \tag{2.2}$$

Finally, a dual representation of the Gauge function of a polyhedral C-set P can be given in terms of its vertex set $V = vert\{P\}$ as

$$\Psi_P(x) = \min\{\mu : x = V\alpha, \; \bar{1}^T \alpha = \mu, \; 0 \le \alpha \in \mathbb{R}^r\}.$$

3. Systems under consideration and problem statement

In this section we will introduce the class of systems under consideration and we will state our problem. Then some considerations concerning the choice of the stabilizing control law will be reported. We will consider the class of continuous and discrete-time uncertain dynamic systems described respectively by

$$\dot{x}(t) = A(w(t))x(t) + B(w(t))u(t) \tag{3.1}$$

$$x(k+1) = A(w(k))x(k) + B(w(k))u(k) \tag{3.2}$$

where the state vector x belongs to $\mathbb{R}^n$, the control values are constrained to belong to a given polyhedral C-set $U \subseteq \mathbb{R}^q$ for every $t \ge 0$. The uncertain state matrices belong to the polytopes of matrices

$$A(w) = \sum_{i=1}^p w_i A_i, \quad B(w) = \sum_{i=1}^p w_i B_i, \tag{3.3}$$

with

$$w \in W = \{w : \sum_{i=1}^p w_i = 1, \;\; w_i \ge 0\}, \tag{3.4}$$

where the given vertex matrices A_i, B_i have appropriate dimensions. In the continuous-time case we will furthermore assume the uncertain function $w(t)$ to be piecewise continuous.

In most practical cases the state is also constrained to belong to a C-set X, for example for safety considerations or for linear model validation, and this implies severe restrictions on the choice of the initial conditions and admissible control laws. To the light of this the problem we will focus on is the following:

Problem 3.1. Given the system (3.1) (respectively (3.2)) with the state constrained to belong to the C-set X for every $t \geq 0$, determine a set $X_0 \subseteq X$ and a stabilizing feedback control law $u(x(t)) = \Phi(x(t))$ such that for every initial condition $x_0 \in X_0$ the closed loop evolution is such that:

$$u(t) \in U \text{ and } x(t) \in X, \quad \forall t \geq 0$$
$$\lim_{t \to +\infty} x(t) = 0.$$

A first solution to this problem can be obviously found by selecting a stabilizing linear static state feedback control law $u = Kx$ (for example proceeding along the lines of [2]) and by picking as set of initial states the ellipsoidal region $X_0 = \{x : x^T P x \leq d\}$, where P is chosen in a way such that its derivative is decreasing along the closed loop system trajectories, and $d \geq 0$ is the maximal value such that $X_0 \subset (X_U \bigcap X)$, being $X_U = \{x : Kx \in U\}$.

Unfortunately an inappropriate choice of the gain K might result in a very small set of attraction whereas we are normally interested in determining, given the constraint sets X and U, an as large as possible, according to some criterion, set of initial states which can be asymptotically driven to the origin. Moreover it is possible to furnish examples of dynamic uncertain linear systems which do not admit any ellipsoidal invariant set while they do admit polyhedral invariant sets.

In the next section we will recall some of the results concerning set-induced Lyapunov functions for the robust stabilization of uncertain linear dynamic systems in the presence of control and state constraints. We will see how a complete solution to Problem 3.1 can be given by determining a proper polyhedral region of attraction which will result in being a Lyapunov function for the system under consideration.

4. Set induced polyhedral Lyapunov functions

The problem definition, in its actual form, does not shed much light on the choice of the feedback control law. Indeed an alternative and constructive way of proceeding (in the sense that it will provide us with the requested control law) is that of reformulating the problem under consideration by means of Lyapunov functions. To this aim we introduce the following definitions.

Definition 4.1. *The C-set $S \subset X$ is a domain of attraction (with speed of convergence β) for system (3.1) if there exists $\beta > 0$ and a continuous feedback control law $u(t) = \Phi(x(t))$ such that for all $x_0 \in S$ the closed loop trajectory $x(t)$ with initial condition $x(0) = x_0$ is such that $u(t) \in U$ for every $t \geq 0$ and*

$$\Psi_S\left(x(t)\right) \leq e^{-\beta t}\Psi_S\left(x(0)\right) \tag{4.1}$$

for every possible piecewise-continuous disturbance $w(t) \in W$. If we take $\beta = 0$, the set S is simply said to be U-invariant [15].

Definition 4.2. *The C-set $S \subset X$ is a domain of attraction (or λ-contractive set) for system (3.2) if there exists $0 \leq \lambda < 1$ and a feedback control law $u(k) = \Phi(x(k))$ such that for all $x_0 \in S$ the closed loop trajectory $x(k)$ with initial condition $x(0) = x_0$ is such that $u(k) \in U$ for every $k \geq 0$ and*

$$\Psi_S\left(x(k)\right) \leq \lambda^k\Psi_S\left(x(0)\right) \tag{4.2}$$

for every possible disturbance sequence $w(k) \in W$.

From the above definition we have that a solution of Problem 3.1 can be found by determining a domain of attraction (which results in being the searched set of initial states X_0 in Problem 3.1) and which is nothing but the unitary level surface of a Lyapunov function for the closed loop system. A better insight in the determination of such domains and in the searched feedback control law can be obtained if we reformulate the decreasing conditions (4.1) and (4.2) in differential terms as done below.

Fact 4.1. *The set $S \subset X$ is a domain of attraction for system (3.1) with speed of convergence β or, respectively, a λ-contractive set for system (3.2) if and only if for every $x \in \partial S$ there exists $u \in U$ such that*

$$D^+\left(x,u\right) \doteq \limsup_{\tau \to 0} \frac{\Psi_S\left(x + \tau(A(w)x + B(w)u)\right) - \Psi_S\left(x\right)}{\tau} \leq -\beta\Psi_S\left(x\right) = -\beta. \tag{4.3}$$

or, in the discrete-time case,

$$\Psi_S\left(A(w)x + B(w)u\right) \leq \lambda\Psi_S\left(x\right) = \lambda. \tag{4.4}$$

holds for every possible $w \in \partial W$.

By exploiting the decreasing conditions just reported it is possible to show that the domains of attraction with a given speed of convergence β (λ) included in a C-set X are subject to a partial ordering in the sense that given two contractive sets S_1 and S_2 their convex hull is also contractive so that it is possible to define the 'largest' contractive set contained in X, thus furnishing us the most natural criterion for the choice of the searched region. This is reported in the next Lemma which, for the sake of clarity, considers only the discrete-time case.

Lemma 4.1. *Assume S_1 and S_1 are two λ-contractive sets contained in a C-set X for system (3.2). Then their convex hull is also λ-contractive.*

Proof. *For every $x \in S = conv(S_1, S_2)$ there must exist a control value u such that*

$$A(w)x + B(w)u \in \lambda S \text{ for every } w \in W.$$

By writing x in terms of $x_1 \in S_1$ and $x_2 \in S_2$ the above can be rewritten as

$$y = \alpha \left[A(w)x_1 + B(w)u_1 \right] + (1 - \alpha) \left[A(w)x_2 + B(w)u_2 \right] \in \lambda S.$$

By definition we can find $u_i \in U$, $i = 1..2$ such that each of the square bracketed terms belongs to $\lambda S_i, i = 1..2$ hence setting

$$u(x) = \alpha u(x_1) + (1 - \alpha)u(x_2)$$

we have that $u(x) \in U$ (being U a C-set) and $y \in \lambda S$ thus proving our assertion.   □

So far we have not imposed any restriction on the class of contractive sets. Fortunately enough, according to the next theorem, we can focus our attention just on the class of polyhedral sets.

Theorem 4.1. *[6] Assume that there exists a β_1-contractive, $\beta_1 > 0$, (λ_1-contractive, $0 < \lambda_1 < 1$) set P_1 contained in the C-set X for the system (3.1) or (3.2). Then, for all β such that $0 < \beta < \beta_1$ (λ such that $\lambda_1 < \lambda < 1$) there exists a polyhedral β-contractive (λ-contractive set) P contained in X.*

This property is known as universality property of the polyhedral sets and allow us to look for a solution to the constrained stabilization problem in this class. In fact from Theorem 4.1 we have that a Lyapunov function exists, say a solution of problem 3.1 exists, if and only if there exists a polyhedral one.

For a polyhedral set we can rephrase the differential conditions (4.3) and (4.4) in terms of a finite number of inequalities. Furthermore, due to the linearity of the constraints and of the systems under consideration, we can just check these inequalities on a finite number of points (instead of the whole border of the region) corresponding to the vertices of the region under observation as summarized next [6].

Theorem 4.2. *The polyhedral C-set $P \subseteq X$ is a domain of attraction for system (3.2) if for every $x \in vert\{P\}$ there exists $u \in U$ such that*

$$\max_i \max_j \{F_i(A_j x + B_j u)\} \leq \lambda \qquad (4.5)$$

The same property holds also in the continuous-time if we replace condition (4.5) with the following subtangentiality condition (note the first subscript)

$$\max_{i \in I(x)} \max_j \{F_i(A_j x + B_j u)\} \leq -\beta \qquad (4.6)$$

The geometric meaning of (4.6) is straightforward: a polyhedral region P is a λ contractive domain for system (3.2) if for each of his vertices there exists a control value $u \in U$ which maps this vertex inside the scaled region λP for every possible disturbance $w \in W$. In the continuous time-case (4.6) requires the existence of a $u \in U$ which directs the velocity vector $A(w)x + B(w)u$ towards the interior of the region for every possible disturbance $w \in W$. The final and major advantage derives from the fact that it is possible to furnish a numerical procedure, which has its basis in the condition we have just reported, to compute the maximal λ-contractive set contained in a given polyhedral C-set X for a discrete-time system. As a final remark we point out that the above results can be expressed in dual terms by means of matrix equalities and inequalities as reported in the next theorems whose proof is given in [10].

Theorem 4.3. *Let $P \subseteq X$ be a C-set and $V = \{v_1, .., v_k\}$ be its vertex representation. Then P is a λ-contractive set for system (3.2) if and only if there exists a set of p nonnegative square matrices $H_i \in \mathbb{R}^{k \times k}$ and a matrix R such that:*

$$A_j V + B_j R = V H_j$$
$$H_j \bar{1} \leq \lambda \bar{1}$$

and each column r_i of the matrix R is such that $r_i \in U$.

In the continuous-time case an equivalent result, which extends the result reported in [13, 25], holds.

Let an $\mathcal{M}$-matrix H be a real matrix such that

$$H_{ij} \geq 0, \text{ if } i \neq j.$$

Theorem 4.4. *Let $P \subseteq X$ be a C-set and R be its vertex representation. Then P is a β-contractive set for system (3.1) if and only there exists a set of p $\mathcal{M}$-matrices H_i and a matrix R such that*

$$A_j V + B_j R = V H_j$$
$$H_j \bar{1} \leq -\beta \bar{1}$$

and each column r_i of the matrix R is such that $r_i \in U$.

5. Construction of a contractive set: the discrete-time case

We now report some results concerning the construction of the maximal λ-contractive set for a discrete-time constrained system as (3.2) and a numerical procedure, which is based on the notion of controllability regions, to generate such a set. By letting the contractivity factor λ tend to 1 this will also allow us to provide an arbitrary good approximation of the maximal set of states which can be brought to the origin while not violating the imposed constraints on the control and state variables.

Given the polyhedral C-sets X and U, the maximal λ-contractive set P_λ contained in X can be constructed starting from $X^0 = X$ by calculating at every step the set X^{k+1} of states which can be mapped through some $u \in U$, for every possible $w \in W$, in the set λX^k, as formalized in the following algorithm.

Algorithm 5.1. Construction of a polyhedral Lyapunov function [5].

Let $X^{(0)} = \{x \in \mathbb{R}^n : F^{(0)}x \leq \bar{1}\} = \{x = V^{(0)}\alpha,\ \bar{1}^T\alpha = 1,\ \alpha \geq 0\}$ be the assigned polyhedral C-set. Fix $\lambda :\ 0 \leq \lambda < 1$, and a tolerance $\epsilon > 0$ such that $\lambda + \epsilon < 1$. Set $k = 0$.

1. Consider the set $Q^{(k)} = \{(x, u) : F^{(k)}[A_i x + B_i u] \leq \lambda\bar{1},\ i = 1, \ldots, p\} \subset \mathbb{R}^{n+q}$
2. Compute the projection of $Q^{(k)}$ on $\mathbb{R}^n :\ P^{(k)} = \{x :\ \exists u \in U :\ (x, u) \in Q^{(k)}\}$;
3. Compute the polyhedron $X^{(k+1)} = X^{(k)} \cap P^{(k)}$;
4. If $0 \in int\{X^{(k+1)}\}$, that is $X^{(k+1)}$ is a C-set, continue, otherwise stop (the procedure has failed with the assigned λ);
   set $X^{(k+1)} = \{x :\ F^{(k+1)}x \leq \bar{1}\} = \{x = V^{(k+1)}\alpha,\ \bar{1}^T\alpha = 1,\ \alpha \geq 0\}$;
5. If $X^{(k+1)}$ is $(\lambda + \epsilon)$-contractive then set $P = X^{(k+1)}$ and stop, else set $k = k + 1$ and go to (i).

For an efficient implementation of the procedure both vertex and plane representation of the initial set X and the sets $X^{(k)}$ are considered [10]. The first element of the pair (x, u) in step (ii) is a state which is mapped by the control $u \in U$ in $\lambda X^{(k)}$ for all $w \in W$. Thus the resulting set is the constrained controllability set to $\lambda X^{(k)}$. The projection operation can be performed via Fourier–Motzkin elimination method (see for instance the work [18]). The set $X^{(k+1)}$ in step (iii) is the subset of all states in $X^{(k)}$ which can be re-mapped in $\lambda X^{(k)}$ itself. Finding a *minimal* representation for this set is the most delicate and time consuming step of the whole procedure. To perform this step we set $\tilde{X} = X^{(k)}$ and we consider the plane description of $P^{(k)} = \{G^{(k)}x \leq \bar{1}\}$ and we intersect the set $\tilde{X}$ with all the half-spaces $\Pi_i^{(k)} = \{x :\ G_i^{(k)}x \leq 1\}$ which generate $P^{(k)}$, say $\tilde{X} := \tilde{X} \cap \Pi_i^{(k)}$, for $i = 1, 2, \ldots$. The intersection of the region $\tilde{X}$ described in terms of vertex matrix $\tilde{V}$ and plane matrix $\tilde{F}$ with the hyper-plane $fx \leq 1$ is computed as follows:

– Compute $\tilde{V}_{ext}, \tilde{V}_{int}$ and $\tilde{V}_{pla}$, the subset of vertices of $\tilde{V}$ which satisfy $fx > 1$, $fx < 1$ and respectively $fx = 1$ (within a certain numerical tolerance).

- If $\tilde{V}_{ext} = \emptyset$ then stop: the result of the intersection is $\tilde{X}$ itself.
- If $\tilde{V}_{int} = \emptyset$ then stop: the result of the intersection is an empty set.
- If $\tilde{V}_{ext} \neq \emptyset$ then discard all the planes of $\tilde{X}$ which don't contain internal vertices, store those which contain internal and external vertices in a matrix $\tilde{F}_{new}$ and store the remaining ones in $\tilde{F}$.
- Determine the set of all the points, which can be obtained by solving the linear equation constituted by considering $fx = 1$ and n-1 planes of $\tilde{F}_{new}$, and store them in a matrix $\tilde{V}_{pos}$.
- Remove from $\tilde{V}_{pos}$ the points which don't satisfy all the constraints induced by $\tilde{F}_{new}$.
- The result of the intersection is the polyhedron whose vertex and plane representation is given by $\tilde{X} = [\tilde{V}_{pos}|\tilde{V}_{int}|\tilde{V}_{pla}]$ and $\tilde{F} = [f^T|\tilde{F}_{new}^T|\tilde{F}^T]^T$.

To check the contractivity of the set P, as required in step (v), we must have that for all the vertices x_i of $X^{(k+1)}$ the optimal values of the LP problems

$$\min \tilde{\lambda} \ \ s.t.$$
$$F^{(k+1)}B_1 u - \tilde{\lambda}\tilde{1} \ \leq \ -F^{(k+1)}A_1 x_i$$
$$\dots$$
$$F^{(k+1)}B_p u - \tilde{\lambda}\tilde{1} \ \leq \ -F^{(k+1)}A_p x_i$$
$$u \in U$$

is less or equal than $\lambda + \epsilon$.

We remark that the last constraint, being the set U polyhedral, can be expressed by means on linear inequalities and that, in view of the high number of inequalities involved, it is normally easier to solve the dual problem. If the procedure stops successfully, the final set P_λ is the largest λ-contractive set in X (up to the tolerance ϵ). As a final remark we have that as $\lambda \to 1^-$ the evaluated set approaches the set of all the states which can be driven to the origin. It is moreover worth recalling [9] that in the absence of control this procedure provides the largest λ-contractive set for the autonomous system

$$x(k+1) = A(w)x(k).$$

6. The continuous-time case

In the continuous-time case condition (4.6) furnishes a necessary and sufficient condition which might be tested to check if a given polyhedral region is β-contractive for (3.1) but, in its actual form, doesn't provide us with any constructive criterion to determine the maximal β-contractive set for system (3.1). The problem we are analyzing in this section can hence be stated in the following way: is it possible to provide a constructive procedure to determine, or at least approximate, the largest β-contractive set for system (3.1)? This question has an affirmative answer and in this section we report the results contained in [7] which allow to provide an arbitrarily close approximation of the maximal domain of attraction by exploiting the relation existing between the original system (3.1) and the discrete-time Euler Approximating System, which is defined next.

Definition 6.1. *Given the continuous -time system (3.1) and* $\tau > 0$ *the Euler Approximating System (EAS) is the following discrete-time system*

$$x(k+1) = [I + \tau A(w(k))]x(k) + \tau B(w(k))u(k), \qquad (6.1)$$

After this simple definition we now summarize the main results concerning the constrained control of a continuous-time dynamic systems. We refer the reader to [6] [7] for the proofs which are omitted here for brevity. The first result we are reporting establishes a close link between the contractive sets of a continuous-time system and those associated to its EAS.

Fact 6.1. If there exists a C-set S which is a domain of attraction for (3.1) with a speed of convergence $\beta > 0$ then for all $\beta' < \beta$ there exists $\tau > 0$ such that the set S is contractive for the EAS with $\lambda' = 1 - \tau\beta'$. Conversely, if for some $0 \leq \lambda < 1$, there exists a λ-contractive C-set P for the EAS (6.1) then P is a domain of attraction for (3.1) with $\beta = \frac{1-\lambda}{\tau}$.

The second main result allows us to determine an 'as close as possible' approximation of the maximal β-contractive set for the continuous-time system under consideration. Thus the two results (the one just reported and the next) provide a complete solution to the constrained stabilization problem.

Fact 6.2. For every $\epsilon_1, \epsilon_2 > 0$ the set $S_{\bar{\beta}}$ (the largest domain of attraction in X for (3.1) with speed of convergence $\bar{\beta} > 0$) can always be internally approximated by a polyhedral C-set P such that $(1 - \epsilon_1)S_{\bar{\beta}} \subset P \subset S_{\bar{\beta}}$ and such that P is a domain of attraction for (3.1) with speed of convergence β, with $\bar{\beta} - \epsilon_2 < \beta < \bar{\beta}$.

In [7] it is actually shown how to derive an internal approximating set P such that $(1 - \epsilon_1)S_{\bar{\beta}} \subset P \subset X$. However it is tedious but not difficult to extend those results to show that it is possible to find a polyhedral set P such that $(1 - \epsilon_1)S_{\bar{\beta}} \subset P \subset S_{\bar{\beta}}$. We skip this point because the first inclusion is the most important.

In simpler words the meaning of the above results is the following: for a given $\beta > 0$ we can get an arbitrarily close approximation of the largest domain of attraction (with speed of convergence β) for (3.1) by applying the numerical procedure reported in the previous section to the EAS (6.1), for an appropriate choice of the parameter τ which finally depends mainly on how close we want this approximation to be. Before going on to the determination of a stabilizing control law for the systems under examination, we would like to let the reader note two things:

– The statement in fact 6.1 can be proved to be false if the exponential approximation (instead of the EAS) is used to compute the maximal β-contractive set. It can in fact be seen by very simple examples [7] that the maximal invariant set for the exponential approximation is not contained in maximal invariant set for the continuous time system.

– If we let $B = 0$ and we assume that the system is stable this result allow us to compute an arbitrarily good approximation of the largest domain of attraction.

7. Linear variable structure and discontinuous control law

Once a polyhedral approximation of the domain of attraction for (3.1) or (3.2) with a certain speed of convergence has been found, a feedback control law has to be provided. In this section we will see how it is possible to derive a linear variable structure stabilizing control law proposed in [15] and we will furnish a procedure for its determination. Then we will present a discontinuous control law which is applicable only to continuous-time single input systems.

Let then
$$P = \{x : F_i x \leq \bar{1}, \ i = 1, \ldots, s\}$$

be the contractive set which resulted from applying the mentioned procedure to the EAS of (3.1) or to (3.2). To each vertex v_i of P remains associated a control value u_i (actually provided by the procedure itself). The set P can be partitioned in simplicial sectors S_h each delimited by the origin and n vertices $v_{h_1} \ldots v_{h_n}$ which lay on the same face $cx = 1$ delimiting the set P (hence the row vector c is either F_i or $-F_i$, for some i) [15]. The partition can me made in such a way that two of these sectors have intersection with empty interior, and the union of all the sectors is P. Thus every $x \in P$ belongs necessarily to at least one of these sectors so that it is possible to define a mapping $h = h(x)$ from the state space to the set of the sectors indexes, where $h(x)$ is the index of a sector containing x. Now, if to each of these sectors we associate the gains K^h defined as
$$K^h = [u_{h_1} \ldots u_{h_n}][v_{h_1} \ldots v_{h_n}]^{-1}$$

where the matrix $[v_{h_1} \ldots v_{h_n}]$ is invertible by construction, we have that the linear variable structure control law defined as
$$u_{vs}(x) = K^{h(x)} x$$

stabilizes the system and guarantees condition (4.1) for every initial state $x(0) \in P$. Moreover it can be shown [6] that this control law is continuous.

It is quite obvious that such partition, to reduce the computational load of the proposed control law, should be computed off line so that on-line the algorithm has just to check to which sector the state belongs (if a state belongs to more than one sector then, in view of the continuity of the control law, any of the gains of the concurring sectors will produce the same control value). If the contractive set under consideration is simplicial, say every face contains exactly n vertices, the determination of the partition is immediate and the required map $h(x)$ can be chosen as any map satisfying $h(x) \in I(x)$, where $I(x)$ is defined in (2.2) (the set $I(x)$ can in fact contain more than one index when x belongs to the boundary of a sector; in this case each gain of the concurring sectors produces the same control value in view of the continuity of the proposed control law u_{vs}). Unfortunately the construction of the maximal λ-contractive set most often generates a non-simplicial set $P \subset \mathbb{R}^n$ (say a set whose delimiting planes contain more than n vertices, see [14]) hence such a partition results in being an essential point for the practical implementation of the controller (apart obviously from the two dimensional case where every polyhedron is simplicial). One possible way to derive the proceeding partition is by means of a 'stretching' procedure which, starting from the original polyhedron P generates a supporting polyhedron $\tilde{P}$ [14] which is nothing but a polyhedral representation of the mentioned partition, here briefly reported. The main idea of the procedure is to supply the polyhedron $\tilde{P}$ having a vertex representation of the form $\tilde{V} = [\mu_1 v_1, \mu_2 v_2, \ldots, \mu_r v_r]$, where $V = [v_1, v_2, \ldots, v_r]$ is the vertex representation of P and the $\mu_i \geq 1$ are proper scaling coefficients chosen in such a way the new polytope $\tilde{P} = conv(\tilde{V})$ is simplicial. In this way, each face of $\tilde{P}$ is associated to the simplicial cone generated by its n vertices, which is in turn associated to a simplicial sector of the original polytope.

Algorithm 7.1. Stretching algorithm [14]
Set $k = 0$ and $\tilde{P}^{(k)} = P$. Label each plane and vertex of $\tilde{P}^{(k)}$ with consecutive numbers and let $I_p(k) = \{1 \ldots s(k)\}$ be the set of indexes of the planes and let $I_v = \{1 \ldots r\}$ be the set of the indexes corresponding to the vertices of $\tilde{P}^{(k)}$.

1. For every $i \in I_p(k)$ create the incidence list $Adj(i) = \{j : F_i^{(k)} v_j^{(k)} = 1\}$ (i.e. of the set of all vertices incident in the i-th plane).
2. If all the adjacency lists contain less then $n + 1$ elements stop (the polyhedron is simplicial) otherwise
3. Pick the first adjacency list which contains more than n indices and pick a vertex $v_j^{(k)}$ from this list
4. Compute the maximum factor μ by which the chosen vertex can be stretched while assuring that

$$vert\{conv\{v_1^{(k)}, \ldots, \mu v_j^{(k)}, \ldots, v_r^{(k)}\}\} = vert\{P^{(k)}\}.$$

5. Set $\bar{v}_j = (1 + \frac{\mu}{2})v_j^{(k)}$
6. Set $vert\{\tilde{P}^{(k+1)}\} = \{v_1^{(k)}, \ldots, \bar{v}_j, \ldots, v_r^{(k)}\}$.

7. Compute the plane representation $\tilde{P}^{(k+1)} = \{F^{(k+1)}x \leq 1\}$ and $I_p(k+1)$
8. Set $k = k + 1$ and go to (1)

The above procedure stops in a finite number of steps and furnishes a simplicial supporting polyhedron $\tilde{P} = \{x : \tilde{F}x \leq \tilde{1}\}$ with the same number of vertices of the original one. In this case the required mapping is given by $h(x) = argmax_i\ \tilde{F}_ix$. We would like to remark two main aspects. The first is that the new polytope $\tilde{P}$ is not contractive in general. It is just an auxiliary polytope whose plane representation allows for the computation of the function $h(x)$. The second comes from the fact that the number of sectors (say the complexity of the compensator) grows up very rapidly as the system dimension increases.

We present now an heuristic but efficient technique which can be normally used to speed up the computation of the supporting polyhedron is that of stretching each of the vertices of the original polyhedron P so that the stretched vertices lay on the surface of a casually generated hyper-ellipsoid containing P itself. This can be easily done in the following way

Algorithm 7.2. Heuristic stretching procedure

1. Generate randomly the elements of an $n \times n$ real matrix Q.
2. Set $S = QQ^T + \mu I$ where I is the $n \times n$ identity matrix and $\mu > 0$ is a parameter which assures that $S > 0$ (say $\Psi(x) = x'Sx > 0$ for every $x \neq 0$).
3. Pick a constant k such that $\Psi(v_i) \leq k$ for every $v_i \in vert\{P\}$.
4. For every $v_i \in vert\{P\}$ set $\tilde{v}_i = v_i\sqrt{\frac{k}{\Psi(v_i)}}$

In most of the cases the authors have seen that this results in being a simplicial polyhedron. It is clear that this procedure should be used as preliminary "polishing" for Procedure 7.1.

To avoid the burdens deriving from the high number of simplicial sectors involved in the on-line computation of the required linear gain the authors have recently proposed a discontinuous control law applicable to single input continuous-time systems and which relies solely on the contractive region P and which is now reported. We let the reader know that passing from continuous to discontinuous control we must pay attention to some issues. The first is that we must assure the existence of the solution. The second is that although we are considering a control which is discontinuous we assume the existence of a continuous control as in definition 4.1.

Suppose a β-contractive region $P = \{x : Fx \leq \tilde{1}\}$ for a single input continuous-time system has been found and that the control constraint C-set can be written as $U = [u_{min}, u_{max}]$. Define the mapping

$$\bar{I}(x) = \min_{i \in I(x)} i$$

which associates (arbitrarily) to every $x \in P$ a single index corresponding to a sector of P and for every x consider the following min-max problem:

$$\nu = \min_{u \in U} \max_{w \in W} F_{\bar{I}(x)}(A(w)x + B(w)u)$$

and let $u_{\bar{I}(x)}$ be the control value for which the minimum is reached. Being the above linear in all its terms it is quite clear that this value is either u_{min} or u_{max} (or the value 0 if there is more than one minimizer). In this way the control law $u(x) = u_{\bar{I}(x)}$ remains defined on the whole state space (this is actually the main reason for the introduction of the mapping $\bar{I}(x)$) and can be proved to be stabilizing as reported in the next result.

Theorem 7.1. *Suppose a polyhedral set P is a β-contractive domain for a single input continuous time-system as in (3.1). Then the discontinuous control law*

$$u(x) = u_{\bar{I}(x)}$$

is such that

$$\Psi_P(x) \leq e^{-\beta t} \Psi_P(x(0))$$

for every initial state $x(0) \in P$.

The proof of the above theorem needs to be supported by the notion of equivalent control [24] and the reader is referred to [7] for details. As a final comment we would like to point out that the cited control law is suitable to handle the case of quantized control devices (see [7]).

8. Application of the control to the two tank system

The system we considered is a laboratory two-tank plant whose structure is that reported in the scheme in figure 8.1.

It is formed by the electric pump EP whose job is that of supplying water to the two parallel pipes P1 and P2 whose flow can be either 0 or U_{max} and is regulated by two on-off electro-valves EV1 and EV2 which are commanded by the signals coming from the digital board BRD1 (not reported in figure 8.1). The two parallel pipes bring water to the first tank T1 which is connected, through P12, to an identical tank T2 positioned at a lower level. From T2 the water flows out to the recirculation basin BA. The two identical variable inductance devices VID1 and VID2, together with a demodulating circuit in BRD1, allow the computer to acquire the water levels of the two tanks. These levels are the state variables of the system.

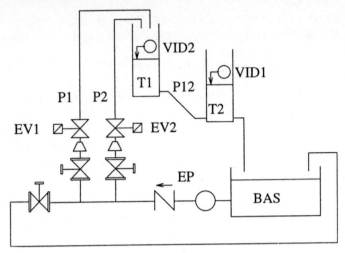

Fig. 8.1. Plant schematic representation

If we denote by h_1 and h_2 the water levels of the two tanks and we choose as linearization point the steady state value $[h_{10} \ h_{20}]^T$ corresponding to the constant input $u_0 = U_{max}$ and we set $x_1(t) = h_1(t) - h_{10}(t)$ and $x_2(t) = h_2(t) - h_{20}(t)$, we get the linearized time-invariant system

$$\begin{bmatrix} \dot{x}_1 \\ \dot{x}_2 \end{bmatrix} = \begin{bmatrix} -\dfrac{\alpha}{2\sqrt{h_{10}-h_{20}}} & \dfrac{\alpha}{2\sqrt{h_{10}-h_{20}}} \\ \dfrac{\alpha}{2\sqrt{h_{10}-h_{20}}} & -\dfrac{\alpha}{2\sqrt{h_{10}-h_{20}}} - \dfrac{\beta}{2\sqrt{h_{20}}} \end{bmatrix} \begin{bmatrix} x_1 \\ x_2 \end{bmatrix} + \begin{bmatrix} 1 \\ 0 \end{bmatrix} u \quad (8.1)$$

where the parameters entering the above equations are $\alpha = .08409$, $\beta = .04711$, $h_{10} = .5274$, $h_{20} = .4014$ and $u_0 = .02985$. To keep into account the effects due to the non linear part of the system we considered the uncertain system described by

$$A(\xi, \eta) = \begin{bmatrix} -\xi & \xi \\ \xi & -(\xi + \eta) \end{bmatrix} \qquad B(\xi, \eta) = \begin{bmatrix} 1 \\ 0 \end{bmatrix} \qquad (8.2)$$

with

$$\xi = .118 \pm .05$$
$$\eta = .038 \pm .01.$$

The state and control constraint sets we considered are respectively given by $X = \{[x_1 x_2]^T : |x_1| \le .1, |x_2| \le .1\}$ and $U = \{-U_{max}, U_{max}\}$.

Starting from X we computed the maximal .2-contractive region by using the corresponding EAS with $\tau = 1$ and $\lambda = .8$. The region representation is given by $P = \{x : |Fx| \le 1\}$ where

$$F = \begin{bmatrix} 1.000 & 0.000 \\ -0.1299 & -1.727 \\ -0.2842 & -1.871 \\ -0.4429 & -1.932 \\ -0.5833 & -1.905 \\ -0.6903 & -1.806 \\ -0.8258 & -1.671 \\ -0.8716 & -1.557 \\ -0.9236 & -1.414 \\ -0.9295 & -1.317 \end{bmatrix} \qquad (8.3)$$

and it is ordered in a way such that each row i of F delimits the sector i according to figure 8.2. This region is formed by 20 symmetric sectors and (as

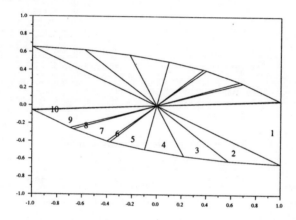

Fig. 8.2. The maximal β-contractive region, with $\beta=0.2$

it is always the case in two dimensions) is simplicial. Hence the computation of the sector gains is immediate and results in 10 different gains which are reported in the matrix K

$$K = \begin{bmatrix} -0.2839 & -0.3003 \\ -1.035 & -1.449 \\ -0.0855 & -0.5613 \\ -0.1329 & -0.5796 \\ -0.1750 & -0.5713 \\ -0.2071 & -0.5419 \\ -0.2477 & -0.5012 \\ -0.2614 & -0.4672 \\ -0.2771 & -0.4243 \\ -0.2788 & -0.3964 \end{bmatrix} \qquad (8.4)$$

which again is ordered in a way such that the i-th row of K corresponds to the i-th sector of P. The result of the implementation of the variable structure control law

$$u(x) = K^{I(x)}x$$

is reported in figure 8.3. We let the reader note that in this simple experiment

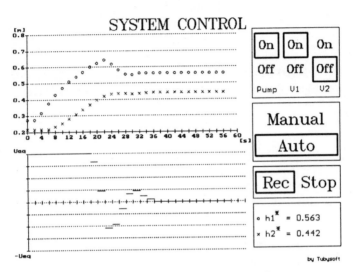

Fig. 8.3. Variable structure control

we didn't force the initial state to belong to the set P. This can be immediately seen from the fact that the control saturates for the first 20 seconds. After this period the system is maintained inside the region and converges asymptotically to the steady state value (the origin of the linearized system) with the assigned contractivity speed.

For this same plant we also implemented the discontinuous control law but, as one can see from the experimental results in figure 8.4, due to the extremely low sampling frequency (1 Hz) the system exhibits a limit cycle thus not converging to the origin (this is anyway in accordance with the theory of sliding modes, see [24]).

An interesting heuristic procedure which the authors have seen to produce good results is that of considering a simple linear control law whose gain is obtained by averaging the gains of the just reported variable structure control law. In our case the average gain is given by

$$k = \begin{bmatrix} -.2984 & -.5792 \end{bmatrix}$$

and the maximal β-contractive region, with $\beta = 0.2$, of the closed loop system included in the non-saturation set $X \cap X_U$, where $X_U = \{x : |kx| \leq .3\}$, resulted in the internal region in figure 8.5. Clearly this set is smaller than the

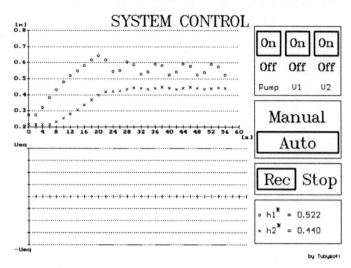

Fig. 8.4. Bang-Bang control system evolution

original one in figure 8.2. However, its existence assures a speed of convergence $\beta = 0.2$ for the closed-loop system with the obtained linear control. In fact the domain of attraction is greater and it is the external region in figure 8.5.

9. Concluding remarks

In this chapter we have reported some recent result concerning the constrained stabilization of linear dynamic systems affected by memoryless uncertainties. We have seen how this problem has a complete solution if formulated in terms of polyhedral Lyapunov functions and we have reported a constructive procedure for the determination of a polyhedral Lyapunov function (if any) for the system under consideration. We have also seen how, starting from the computed polyhedral Lyapunov function, it is possible to derive stabilizing control laws.

Finally the application of the proposed techniques to a laboratory control system has been provided together with some heuristic considerations regarding the possibility of deriving a simple linear control law. This gain can be obtained by averaging those of the various sectors concurring in the determination of the nonlinear control law and in most of the case the authors have seen that this results in a quite good closed behavior.

This point, say the reduction of the complexity of the proposed controller, is an issue which surely needs some further study. Another aspect which is currently under investigation concerns the possibility of deriving, differently from what we have seen here, stabilizing control laws in explicit form. Another important aspect which has not been considered here but plays an important

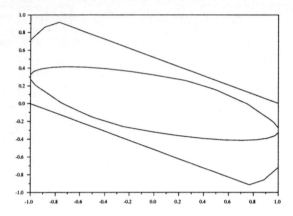

Fig. 8.5. The largest β-contractive and invariant sets with $u = kx$.

role in control theory is certainly that of output feedback stabilization. To the authors' knowledge there are still no necessary and sufficient conditions for the output feedback stabilization of the class of systems considered here.

References

1. A. Benzaouia and C. Burgat, "The regulator problem for a class of linear systems with constrained control", *Syst. and Contr. Letters*, Vol.10, p. 357-363, 1988.
2. J. Bernussou , P.L.D. Peres, and J. Geromel, "On a convex parameter space method for linear control design of uncertain systems", SIAM J. on Contr. and Opt., 29, 381-402, 1991.
3. D.P. Bertsekas, I.B. Rhodes, "On the minmax reachability of target set and target tubes", Automatica, Vol. 7, pp.233-247, 1971.
4. F. Blanchini, "Constrained control for uncertain linear systems", Int. Journ. of Opt. Th. and Appl., Vol. 71, no. 3. p. 465-484, 1991.
5. F. Blanchini, "Ultimate boundedness control for discrete-time uncertain system via set-induced Lyapunov functions", IEEE Trans. on Autom. Contr., Vol. 39, no. 2, 428-433.
6. F. Blanchini, "Nonquadratic Lyapunov functions for robust control", Automatica, Vol. 31, no 3, p. 451-461, 1995.
7. F. Blanchini, S. Miani, "Constrained stabilization of continuous-time linear systems", Systems and Control Letters, Vol. 28, pp. 95-102, 1996.
8. F. Blanchini, S. Miani, "A new class of universal Lyapunov function for the control of uncertain systems", proc. of the CDC96, Kobe, Japan, 1996, to appear.
9. F. Blanchini, S. Miani, "On the transient estimate for linear systems with time-varying uncertain parameters", IEEE Trans. On Circ. and Syst.-I, Vol. 43, no. 7, July 1996.

10. F. Blanchini, S. Miani, "Piecewise-linear functions in robust control", lecture notes in Control and Information Science, Springer-Verlag, New York, 1996.
11. F. Blanchini, F. Mesquine, S, Miani, "Constrained stabilization with assigned initial condition set", Int. Journal of Contr., Vol. 62, no. 3, p. 601-617, 1995.
12. R.K. Brayton, C.H. Tong, "Constructive stability and asymptotic stability of dynamical systems", IEEE Trans. on Circ. and Syst., Vol. CAS-27, no.11, pp. 1121-1130, 1980.
13. E.B. Castelan, J.C. Hennet, "Eigenstructure assignment for state constrained linear continuous-time systems", Automatica, Vol. 28, no. 3, pp. 605-611.
14. B. Grumbaum, "Convex polytopes", John Wiley & Sons, New York, 1967.
15. P.O. Gutman, M. Cwikel, "Admissible sets and feedback control for discrete-time linear systems with bounded control and states", IEEE Trans. on Autom. Contr., Vol. AC-31, No. 4, pp. 373-376, 1986.
16. P.O. Gutman and P.Hagander, "A new design of constrained controllers for linear systems", IEEE Trans. on Autom. Contr., AC-30, 22-33, 1985.
17. A.T. Fuller, In-the-large stability of relay and saturating control systems with linear controllers, Int. J. Control, 10(1969) 457-480.
18. S.S. Keerthi, E.G. Gilbert, "Computation of minimum-time feedback control laws for discrete-time systems with state-control constraints", IEEE Trans. on Autom. Contr., Vol AC-32, no. 5, pp. 432-435, 1987.
19. J.B. Lasserre, "Reachable, controllable sets and stabilizing control of constrained systems", Automatica, Vol. 29, No. 2, pp. 531-536, 1993.
20. H.J. Sussman, E.D. Sontag, and Y. Yang, "A general result on the stabilization of linear systems using bounded controls", IEEE Trans. On Autom. Contr, Vol. AC-39, no.12, pp. 2411-2425, 1994.
21. M. Sznaier, "A set-induced norm approach to the robust control of constrained linear systems", SIAM Journal on Control and Optimization, Vol.31, No.3, 733-746, 1993.
22. K. Tan, E. Gilbert, "Linear systems with state and control constraints: the theory and the applications of the maximal output admissible sets", IEEE Trans. on Autom. Contr., Vol. AC-36, n.9, pp. 1008-1020, 1991.
23. S. Tarbouriech and C. Burgat, "Positively invariant sets for constrained continuous–time systems with cone properties", IEEE Trans. on Autom. Contr., Vol. 39, no. 2, pp.401-405, 1994.
24. I. Utkin, "Sliding modes and their applications in variable structure systems", Moscow, Mir Publisher, 1974.
25. M. Vassilaki, Bitsoris, "Constrained regulation of linear continuous-time dynamical systems", Systems & Control Letters, Vol. 13, 247-252, 1989.

Chapter 7. H_∞ Output Feedback Control with State Constraints

Alexandre Trofino[1], Eugênio B. Castelan[1], and Arão Fischman[2]

[1] Laboratório de Controle e Microinformática (LCMI/EEL/UFSC),
Universidade Federal de Santa Catarina,
PO 476, 88040-900, Florianópolis (S.C.), Brazil.
e-mail: trofino@lcmi.ufsc.br
[2] Laboratoire d'Automatique de Grenoble (URA CNRS 228),
ENSIEG, BP 46, 38402 St.-Martin-d'Hères, France.

1. Introduction

During the last few years the problem of determining output feedback controllers via convex programming techniques has received a lot of attention. The convexity property is an essential feature in optimization problems because, if this property is not present, the existing numerical procedures may lead to a local minimum. This may be a serious drawback since the desired control features, like stability for instance, are garanteed, in general, only at the global optimum.

It has been shown [3], [4] that the solution of several important control problems may be obtained by solving convex (quasiconvex) non differentiable optimization problems. Powerful algorithms and toolbox for solving this type of problems are avalaible to date.

To the authors knowledge the results found in the literature addressing the determination of feedback controllers via convex programming under necessary and sufficient conditions are restricted to the cases of state feedback and dynamic output feedback controllers with the same order of the generalized plant. Design methods for reduced order controllers exhibiting convex properties are unfortunately based on sufficient conditions that may be quite conservative [12].

The main contributions of this paper are necessary and sufficient conditions for the existence as well as a method for the determination of static (and also dynamic) output feedback controllers via biconvex programming. The stability is guaranteed by using a quadratic Lyapunov function.

An interesting feature of the proposed method is that the matrix of output feedback gains appears explicitly in the expressions. In particular, it allows us to show how to combine frequency and time domain specifications as requeriments for the closed-loop system. See, for instance, [15] for further motivation on this type of problem. In this paper we consider H_∞ performance and state constraints in the presence of disturbances. Then, algebraic necessary and sufficient conditions to get the positive invariance property of polyhedral sets in presence of disturbances [2], [6] are derived and used in the optimization program to guarantee the respect of the state constraints.

Notation

Throughout this paper we use the following notation: capital and small letters are used to denote respectively matrices and vectors; $X > Y$ means that $X - Y$ is positive definite; and $X \succeq Y$ (or $x \succeq y$) means that the inequality is component by component, $i.e.$ $X_{ij} \geq Y_{ij}$ for all i, j (or $x_i \geq y_i$ for all i).

2. Problem Statement

Consider the linear time-invariant discrete-time system:

$$x(k+1) = Ax(k) + Bu(k) + B_p p(k) \quad , \quad \forall\, k \geq 0 \qquad (2.1)$$
$$y(k) = Cx(k) \qquad (2.2)$$
$$z(k) = Ex(k) + Fp(k) \qquad (2.3)$$

with $A \in \Re^{n \times n}$, $B \in \Re^{n \times m}$ and $C \in \Re^{q \times n}$. The vectors $p(k) \in \Re^{m_p}$ and $z(k) \in \Re^{q_z}$ define, respectively, a disturbance input and an auxiliary performance output. All matrices are constant and have appropriate dimensions.

Given a matrix G and vectors ξ_l, ξ_u and g, we also consider the two following closed and convex sets:

– admissible disturbance inputs, $\Delta \subset \Re^{m_p}$:

$$\Delta \triangleq \{\, p \in \Re^{m_p} \;;\; -\xi_l \preceq p(k) \preceq \xi_u \,\} \quad \text{with: } \xi_l, \xi_u \succeq 0; \qquad (2.4)$$

– state constraints, $\mathcal{R}[G, g] \subset \Re^n$:

$$\mathcal{R}[G, g] \triangleq \{\, x \in \Re^n \;;\; Gx(k) \preceq g \,\}$$
$$\text{with: } G \in \Re^{r \times n}, \;\; rank(G) = n \;\; \text{and} \;\; g \succeq 0. \qquad (2.5)$$

Mixed H_∞/State constrained control problem

In this paper we are interested in the determination of a stabilizing control law of the type:

$$u(k) = -Ky(k) \quad ; \quad K \in \Re^{m \times q} \qquad (2.6)$$

such that the closed-loop system:

$$x(k+1) = (A - BKC)x(k) + B_p p(k) \quad , \quad \forall\, k \geq 0 \qquad (2.7)$$

satisfies the following two requirements:

- $(r1)$ the norm of the transfer function from $p(k)$ to $z(k)$ satisfies:

$$\| E(zI - A + BKC)^{-1}B_p + F \|_\infty < \gamma \qquad (2.8)$$

- $(r2)$ for any state x_0 belonging to the set $\mathcal{R}[G, g]$ and for any sequence $p(k)$ belonging to Δ, the corresponding state trajectory do not violate the state constraints, $i.e.$ do not leave the region $\mathcal{R}[G, g]$. In other words, $\mathcal{R}[G, g]$ should be a Δ-positively invariant set of the system (2.7) [2], [6].

3. H_∞ results

Consider the linear time invariant discrete time system:

$$\begin{cases} x(k + 1) = Ax(k) + Bu(k) \\ y(k) = Cx(k) \end{cases} \qquad (3.1)$$

with $A \in \Re^{n \times n}$, $B \in \Re^{n \times m}$, $C \in \Re^{q \times n}$.

Now, we are interested in the determination of a static output feedback control law of the type (2.6) that stabilizes the system (3.1).

It is well known that a necessary and sufficient condition for the existence of such a control law is the existence of matrices $P = P^T > 0$ and K such that the following inequality is satisfied:

$$P - (A - BKC)^T P(A - BKC) > 0 \qquad (3.2)$$

Solving the above matrix inequality for K and P is indeed a difficult task because the product of the matrices P and K renders nonconvex the associated numerical problem.

The following result gives necessary and sufficient conditions for the existence of the matrices $P = P^T > 0$ and K satisfying (3.2).

Theorem 3.1. *There exists a pair (P, K), satisfying (3.2) if and only if the solution (P^*, W^*, K^*) of the biconvex optimization problem below, $\mathcal{P}(A, B, C)$, satisfies the condition $P^*W^* = I_n$.*

$$\min_{P,W,K} Tr(PW - I_n) \qquad (3.3)$$

such that:

$$(i) \quad \begin{bmatrix} P & I_n \\ I_n & W \end{bmatrix} \geq 0 \quad ; \quad W > 0$$

$$(ii) \quad \begin{bmatrix} P & (A - BKC)^T \\ (A - BKC) & W \end{bmatrix} > 0$$

Proof:
Necessity: From (3.2), we have the following inequalities satisfied:

$$\begin{cases} P > 0 \\ P - (A - BKC)^T P(A - BKC) > 0 \end{cases} \tag{3.4}$$

Using the Schur complement we can rewrite (3.4) as:

$$\begin{bmatrix} P & (A - BKC)^T \\ (A - BKC) & P^{-1} \end{bmatrix} > 0 \tag{3.5}$$

Defining $P^{-1} \triangleq W$, then (3.5) implies that the conditions (3.3.i)-(3.3.ii) are satisfied and $Tr(PW - I_n) = 0$. Since the objective function is non-negative from (3.3.i), the pair (P, W) is an optimal solution of the problem (3.3).

Sufficiency:. Suppose that the optimal solution of the problem (3.3), denoted (P^*, W^*, K^*), satisfies $(P^* W^* = I_n)$. Thus, $Tr(P^* W^* - I_n) = 0$ and from (3.3.i) we have $W^* = P^{*-1}$. This implies that the condition (3.3.ii) is equivalent to (3.5) which is equivalent to (3.4) and (3.2). *Box*

Remark 3.1. The optimization problem $\mathcal{P}(A, B, C)$ has convex constraints but it is not globally convex because the optimization function is not convex. However, $\mathcal{P}(A, B, C)$ is biconvex in P and W, i.e. it is convex in P for fixed W and in W for fixed P. See [13] for further discussion on these type of problems.

In the sequel we show how theorem 3.1 can be used for the determination of H_∞ controllers for the system (3.1).

Let us consider in (3.1) the auxiliary input vector $p(k) \in \Re^{m_p}$ and the auxiliary performance output vector $z(k) \in \Re^{q_z}$ such that:

$$\begin{cases} x(k+1) = (A - BKC)x(k) + B_p p(k) \\ z(k) = Ex(k) + Fp(k) \end{cases} \tag{3.6}$$

with B_p, E and F being given matrices of appropriate dimensions. Suppose we are interested in the determination of a controller K such that the norm of the transfer function from $p(k)$ to $z(k)$ in (3.6) satisfies:

$$\| E[zI - (A - BKC)]^{-1} B_p + F \|_\infty < \gamma \tag{3.7}$$

Without loss of generality, we assume in the sequel that $q_z = m_p$. Since any rank condition is required on matrices E, F and B_p, this assumption can always be met by completing matrices E and F (or B_p and F) with some null rows (or columns).

A necessary and sufficient condition for (3.7) to be satisfied for some K, is the existence of $P > 0$ and K such that [8]:

$$\begin{bmatrix} A - BKC & B_p \\ E & F \end{bmatrix}^T \begin{bmatrix} P & 0 \\ 0 & \gamma^{-1}I_{q_z} \end{bmatrix} \begin{bmatrix} A - BKC & B_p \\ E & F \end{bmatrix} - \begin{bmatrix} P & 0 \\ 0 & \gamma I_{q_z} \end{bmatrix} < 0$$

$$(3.8)$$

With simple change of variables that issue from the comparison of (3.8) with (3.2) we may use the Theorem 3.1 to find matrices $P > 0$ and K satisfying (3.8) (or declare that (3.8) and equivalently (3.7) have no solution).

If the objective is the minimization of the H_∞ norm of the transfer function in (3.7) then we must apply iteractively theorem 3.1 with smaller values for γ until its minimum value is achieved. Notice that this problem is affine in γ.

4. Mixed H_∞/State Constrained results

As previously quoted, to satisfy requirement $(r2)$ of the problem statement, the controller must be such that the polyhedral set of state constraints, $\mathcal{R}[G, g]$, is a Δ-positively invariant set of the closed-loop system (2.7). An *internal* characterization of the Δ-positive invariance property, in terms of the extremal points of the sets $\mathcal{R}[G, g]$ and Δ, is given in [2]. For our purposes, a more convenient characterization is obtained from an external description of the two polyhedral sets and is considered below.

Proposition 4.1. *A necessary and sufficient condition to get the Δ-positive invariance of $\mathcal{R}[G, g]$ in closed-loop is the existence of an output feedback matrix $K \in \Re^{m \times q}$ and a non-negative matrix $M \in \Re^{r \times r}$ $(M \succeq 0)$ such that:*

$$MG = G(A - BKC) \qquad (4.1)$$

$$Mg + (GB_p)^+ \xi_u + (GB_p)^- \xi_l \preceq g \qquad (4.2)$$

$$where: \begin{cases} (GB_p)_{ij}^+ = max\{(GB_p)_{ij}, 0\} \\ (GB_p)_{ij}^- = max\{-(GB_p)_{ij}, 0\} \end{cases} \quad \forall i = 1, \dots, r \ and \ \forall j = 1, \dots, m$$

Proof:

Necessity: In closed-loop, the Δ-positive invariance property of $\mathcal{R}[G, g]$ can be described as follows:

$$G[A - BKC \quad B_p] \begin{bmatrix} x(k) \\ p(k) \end{bmatrix} \preceq g \ ,$$

$$\forall \begin{bmatrix} x(k) \\ p(k) \end{bmatrix} \text{ such that } \begin{bmatrix} G & 0 \\ 0 & I_{m_p} \\ 0 & -I_{m_p} \end{bmatrix} \begin{bmatrix} x(k) \\ p(k) \end{bmatrix} \preceq \begin{bmatrix} g \\ \xi_u \\ \xi_l \end{bmatrix} \qquad (4.3)$$

Using the extension of Farkas' Lemma presented in [9], (4.3) is satisfied if and only if there exists a matrix $\mathcal{M} = [M \ K_1 \ K_2] \succeq 0$, with $\mathcal{M} \in \Re^{r \times (r + 2m_p)}$, such that:

$$[M\ K_1\ K_2] \begin{bmatrix} G & 0 \\ 0 & I_{m_p} \\ 0 & -I_{m_p} \end{bmatrix} = G[A - BKC\ B_p] \qquad (4.4)$$

$$[M\ K_1\ K_2] \begin{bmatrix} g \\ \xi_u \\ \xi_l \end{bmatrix} \preceq g \qquad (4.5)$$

The first equality (4.1) follows directly from (4.4) which also gives $GB_p = (K_1 - K_2)$. By using the definitions of $(GB_p)^+$ and $(GB_P)^-$ and the fact that $K_1 \succeq 0$ and $K_2 \succeq 0$, we have: $(K_1 - K_2) = (GB_p)^+ - (GB_P)^-$ and $|K_1 - K_2| = (GB_p)^+ + (GB_P)^- \preceq K_1 + K_2$. Together, these last two relations give $(GB_p)^+ \preceq K_1$ and $(GB_P)^- \preceq K_2$, which allow to obtain (4.2) from (4.5).

Sufficiency: Consider that (4.1) and (4.2) are satisfied and let $x(k)$ and $p(k)$ be such that $x(k) \in \mathcal{R}[G, g]$ and $p(k) \in \Delta$. From (4.1), $Gx(k + 1) = G(A - BKC) + GB_p p(k) = MGx(k) + GB_p p(k)$. But, by assumption, $\xi_l \preceq p(k) \preceq \xi_u$, and hence:

$$GB_p p(k) = [(GB_p)^+\ (GB_p)^-] \begin{bmatrix} p(k) \\ -p(k) \end{bmatrix} \preceq [(GB_p)^+\ (GB_p)^-] \begin{bmatrix} \xi_u \\ \xi_l \end{bmatrix}$$

Thus, from (4.2), we get: $Gx(k + 1) = Mg + (GB_p)^+ \xi_u + (GB_p)^- \xi_l \preceq g$.
□

Notice that relations (4.1) and (4.2) reduces to the well known positive invariance relations in the case of unforced systems, that is $B_p = 0$ (see, for instance, [1], [9]). They can also be specialized to the case of symmetric and non symmetric polyhedral sets as well as to consider the case of linear uncertain systems [10]. Notice also that no rank condition on G is required in the proof of the proposition 1 and, hence, relations (4.1) and (4.2) are also valid in the case $rank(G) < n$. However, additional stabilizing conditions related to the kernel of G have to be considered in this case of unbounded polyhedral sets [5] [7].

The external characterization of the Δ-positive invariance property given above is, in general, computationally more attractive than the one give in [2] in terms of the extremal points of the considered polyhedral sets. Furthermore, (4.1) and (4.2) are linear and convex and, as pointed out in [10], [14], [17], they can be transformed into a linear programming problem to achieve a controller satisfying the state constraints as defined in requirement *(r2)*.

In order to achieve both requirements *(r1)* and *(r2)*, we take advantage of the linearity and convexity property of the Δ-positive invariance relations and we propose to consider (4.1) and (4.2) as additional constraints to the H_∞ problem shown in (3.8) (see also [16]). In this way, let us define

$$v \stackrel{\Delta}{=} (GB_p)^+ \xi_u + (GB_p)^- \xi_l$$

and consider matrices $G^\dagger \in \Re^{n \times r}$ and $N \in \Re^{(n-r) \times r}$ such that: $\begin{bmatrix} G^\dagger \\ N \end{bmatrix} G = \begin{bmatrix} I_n \\ 0 \end{bmatrix}$. In order to reduce the number of constraints and variables of the optimization problem related to the Δ-invariance relations, we obtain from (4.1):

$$M = G(A - BKC)G^\dagger + DN$$

where $D \in \Re^{r \times (r-n)}$ is also to be determined. Thus, we get the following optimization problem to solve the *Mixed H_∞/State constrained control problem* defined in section 2:

Proposition 4.2. *There exists a stabilizing output feedback law of the form (2.6) such that the requirements (r1) and (r2) are both satisfied if and only if the solution (P^*, W^*, K^*, D^*) of the optimization problem bellow satisfies the condition $P^*W^* = I$.*

$$\min_{P,W,K,D} Tr(PW - I_n) \tag{4.6}$$

such that:

(i) $\begin{bmatrix} P & I_n \\ I_n & W \end{bmatrix} \geq 0 \quad ; \quad W > 0$

(ii) $\begin{bmatrix} \begin{bmatrix} P & 0 \\ 0 & \gamma I_{q_z} \end{bmatrix} & \begin{bmatrix} (A-BKC) & B_p \\ E & F \end{bmatrix}^T \\ \begin{bmatrix} (A-BKC) & B_p \\ E & F \end{bmatrix} & \begin{bmatrix} W & 0 \\ 0 & \gamma I_{q_z} \end{bmatrix} \end{bmatrix} > 0$

(iii) $GAG^\dagger - GBKCG^\dagger + DN \succeq 0$

(iv) $(GAG^\dagger - GBKCG^\dagger + DN - I)g + v \preceq 0$

Remark 4.1. It is worth noticing that other convex constraints can be introduced into the optimization problem (4.6). In particular, we can also look for a matrix K with pre-specified bounds on its norm or even on its entries, or to obtain, in the case of hard control constraints given by $\{-\rho_l \preceq u(k) \preceq \rho_u\}$, with $(\rho_u, \rho_l) \succeq 0$, admissible controls satisfying $\mathcal{R}[G, g] \subset \mathcal{R}[KC, \rho_u, \rho_l]$, where: $\mathcal{R}[KC, \rho_u, \rho_l] \overset{\triangle}{=} \{ x \in \Re^n \; ; \; -\rho_l \preceq KCx \preceq \rho_u \}$ (see [10], [9]).

5. Numerical Example

In this section we present a numerical example to illustrate the results of Proposition 2.

Let us consider the system in (1),(2),(3) with the following data [10]:

$$A = \begin{bmatrix} -0.0700 & -0.84280 \\ 0.0588 & -0.25425 \end{bmatrix} \quad ; \quad B = \begin{bmatrix} 1 \\ 0 \end{bmatrix} \quad ; \quad Bp = \begin{bmatrix} 0 \\ 1 \end{bmatrix}$$

$$C = \begin{bmatrix} 1 & 0 \end{bmatrix} \quad ; \quad E = \begin{bmatrix} 0 & 1 \end{bmatrix} \quad ; \quad F = 0$$

and state and disturbance constraints (2.4),(2.5) given by:

$$G = \begin{bmatrix} 1 & 0 \\ 0 & 1 \\ -1 & 0 \\ 0 & -1 \end{bmatrix} \quad ; \quad g = \begin{bmatrix} 1 \\ 1 \\ 1 \\ 1 \end{bmatrix} \quad ; \quad \xi_l = \xi_u = 0.4$$

By choosing $\gamma = 1.25$ in (3.7), and solving the optimization problem (4.6), we get the following optimal solution:

$$P = \begin{bmatrix} 0.1936370 & -0.2203618 \\ -0.2203618 & 1.1990283 \end{bmatrix} \quad ; \quad W = \begin{bmatrix} 6.5300567 & 1.2001202 \\ 1.2001202 & 1.0545739 \end{bmatrix}$$

$$K = \begin{bmatrix} 0.0491402 \end{bmatrix} \quad ; \quad D = \begin{bmatrix} 0.6048063 & 0.0935081 \\ 0.2437477 & 0.1249646 \\ 0.6048061 & 0.0935090 \\ 0.2437455 & 0.1249548 \end{bmatrix}$$

6. Concluding Remarks

A design procedure for static output feedback is proposed in this paper. It consists of minimizing a bilinear function subject to LMI's constraints. Several time and frequency domain constraints may be incorporated to the problem, provided they are convex.

An interesting feature is that H_∞, state and control constraints may be jointly treated as shown in proposition 2. Moreover the extension to the case of uncertain systems with polytopic uncertainties is straightforward due to the convex properties of the constraints [4]. The extension to the case of continuous time systems is being investigated.

We emphasize that bilinear functions are unfortunately not globally convex in general and hence the problem may be computationally hard. See [13] for further details, motivations and references on biconvex problems.

References

1. G. Bitsoris, *On the positive invariance of polyhedral sets for discrete-time systems.* Systems and Control Letters, 11, pp.243-248, 1988.
2. F. Blanchini, *Feedback Control for Linear Time-Invariant Systems with State and Control Bounds in the Presence of Disturbances.* IEEE Transactions on Automatic Control, Vol.35, No.11, pp.1231-1234, 1991.

3. S. Boyd and C. Barrat, *Linear Controller Design: Limits of Performance.* Prentice Hall, 1991.

4. S. Boyd, El Ghaoui, E. Feron and V. Balakrishnan, *Linear Matrix Inequalities in Systems and Control Theory,* SIAM, 1994.

5. E.B. Castelan and J.C. Hennet, "Eigenstructure Assignment for State Constrained Continuous Time Systems", *Automatica,* vol. 28, No 3, pp. 605-611, 1992.

6. E. De Santis, *On Positively Invariant Sets for Discrete-Time Linear Systems with Disturbance: An Application of Maximal Disturbance Sets.* IEEE Transactions on Automatic Control, Vol.39, No.1, pp245-249, 1994.

7. C.E.T. Dórea and B.E.A. Milani, "Design of L-Q Regulators for State Constrained Continuous-Time Systems", *IEEE Trans. Automatic Control,* vol. 40, No 3, pp. 544-548,1995.

8. J. Doyle, A. Packard and K. Zhou (1991). *Review of LFTs, LMIs an μ,* Proceedings of the 30th IEEE Conference on Decision and Control, Brighton, December 1991.

9. J.C. Hennet, *Une extension du Lemme de Farkas et son application au problème de régulation linéaire sous contraintes.* C.R. Académie des Sciences, t.308, Série I, pp.415-419, 1989.

10. B.E.A. Milani, E.B. Castelan and S. Tarbouriech, *Linear Regulator Design for Bounded Uncertain Discrete-time Systems with Additive Disturbances.* Proceedings of the 13th World Congress of IFAC (Vol. G), San Francisco, June/July 1996.

11. R.A. Paz and E. Yaz, *Robust Stabilization and Disturbance Attenuation for Discrete-Time Systems with Structured Nonlinearities.* Proceedings of the 32th IEEE Conference on Decision and Control, San Antonio, December 1993.

12. P.L.D. Peres, J.C. Geromel and S.R. Souza, H_2 *Output Feedback Control for Discrete-Time Systems.* Proceedings of the 1994 American Control Conference, Baltimore, June/July 1994.

13. M.G. Safonov, K.C. Goh and J.H. Ly, *Control system synthesis via bilinear matrix inequalities.* Proceedings of the American Control Conference, Baltimore, June 1994.

14. M. Sznaier, *A set induced norm approach to the robust control of constrained systems.* SIAM J. Control and Optimization, Vol. 31, No. 3, pp.733-746, 1993.

15. M. Sznaier and Z. Benzaid, *Robust control of systems under mixed time/frequency domain constraints via convex optimization.* Proceeding of the 31st IEEE Conference on Decision and Control, Tucson, December 1992.

16. A.Trofino Neto, E.B. Castelan and A. Fischman, H_∞ *Control with States Constraints.* Proceedings of th 34th IEEE Conference on Decision and Control, New Orleans, December 1995.

17. M. Vassilaki, J.C. Hennet and G. Bitsoris, *Feedback control of discrete-time systems with state and control constraints.* Int. Journal of Control, Vol. 47, 1988, pp.1727-1735.

Chapter 8. Dynamic Output Feedback Compensation for Systems with Input Saturation

Feng Tyan[1] and Dennis S. Bernstein[2]

[1] Department of Aerospace Engineering TamKang University, Tamsui, Taipei Hsien, Taiwan 25137

[2] Department of Aerospace Engineering The University of Michigan Ann Arbor, MI 48109-2118

1. Introduction

The need for controlling dynamic systems subject to input saturation is a widespread problem in control engineering of immense practical importance. The growing literature on this subject (see, for example, [1, 2, 3, 4, 5, 6, 7, 8, 9, 10, 11, 12, 13, 14]) addresses both stabilization and disturbance rejection objectives subject to constraints on the amplitude of the control input.

In the present paper we use optimization techniques to synthesize feedback controllers that provide local or global stabilization along with suboptimal performance for systems with input saturation. Our approach is based upon LQG-type fixed-structure techniques that yield both full- and reduced-order, linear and nonlinear controllers. This technique was applied to the control saturation problem in [15], where stability and performance bounds were guaranteed by a small-gain-type analysis applied to a deadzone nonlinearity as in [16]. The development in [15] extended the approach of [16] by using a novel Lyapunov-like argument to provide a guaranteed subset of the domain of attraction while allowing the control signal to enter the saturated region.

In the present paper we consider a variation of the approach of [15] by applying positivity-type analysis to guarantee local or global stability. Although the general framework is analogous to that of [15], the positivity conditions for guaranteeing stability have a structure that is distinct from the small-gain conditions. The resulting synthesis equations, which involve coupled Riccati and Lyapunov equations, are thus different from those of [15]. Although the approach of the present paper is related to [15], we note that the present paper is self-contained in the sense that it does not depend upon any of the results given in [15].

A key aspect of the approach of the present paper, as well as [15], is the guaranteed subset of the domain of attraction of the closed-loop system. In related work [10, 11, 12, 13], local or global stability are based upon a priori

[0] This research was supported in part by the Air Force Office of Scientific Research under Grants F49620-92-J-0127 and F49620-95-1-0019.

assumptions that the initial conditions and the states of the system lie in a predefined compact set, and in turn the control input lies in a bounded region. However, our approach does not require such assumptions. In fact, our main result, Theorem 2.1, assumes instead that the initial condition lies in a prescribed region which is a subset of the domain of attraction. The resulting control signal is thus free to saturate during closed-loop operation without loss of stability. The specified subset of the domain of attraction thus provides a guaranteed region of attraction.

While the positivity approach of the present paper is distinct from the small-gain approach of [15], the examples considered herein seem to indicate that the positivity approach is less conservative than the results of [15] in estimating the domain of attraction while yielding controllers requiring smaller input slew rates. In fact, although the control signals in [15] exhibited bang-bang behavior, this behavior has not been observed when using the results of the present paper.

The contents of the paper are as follows. In Section 2 we present the main result (Theorem 2.1) for guaranteeing local or global stability with a guaranteed domain of attraction. This result provides the basis for synthesizing linear controllers in Section 3 that may be of either full or reduced order. In Section 4, we consider full- and reduced-order nonlinear controllers with a classical anti-windup structure. In order to guarantee that the optimization conditions have a consistent structure for accommodating the nonlinear terms, the Lagrange multiplier is chosen to satisfy an appropriate Lyapunov equation given by equation (4..12). An alternative structure given by equation (5..8) is investigated in Section 5 where different equations are given for full- and reduced-order nonlinear dynamic compensator synthesis. In Section 6, we demonstrate an iterative method for solving the coupled matrix equations given in Proposition 3.2. Several numerical examples are given in Section 7 to demonstrate both linear and nonlinear controllers.

Notation

I_r	$r \times r$ identity matrix
$\mathbf{S}^n, \mathbf{N}^n, \mathbf{P}^n$	$n \times n$ symmetric, nonnegative-definite, positive-definite matrices
$\lambda_{\max}(F), \lambda_{\min}(F)$	maximum and minimum eigenvalues of matrix F having real eigenvalues
$\|x\|$	Euclidian norm of x, that is, $\|x\| = \sqrt{x^T x}$
Re	real part
$(\cdot)^*$	complex conjugate transpose

2. Analysis of Systems with Saturation Nonlinearities

In this section, we develop the foundation for our controller synthesis approach given in later sections. Instead of dealing with the saturation nonlinearity directly, we transform the problem in terms of a deadzone nonlinearity as in [16]. The main reason for doing this is the ability to include this nonlinearity in a sector $[0, 1 - \beta_\ell]$, where β_ℓ is a parameter that will be specified in Theorem 2.1. Consider the closed-loop system

$$\dot{\tilde{x}}(t) = \tilde{A}\tilde{x}(t) + \tilde{B}(\sigma(u(t)) - u(t)), \quad \tilde{x}(0) = \tilde{x}_0, \tag{2..1}$$

$$u(t) = \tilde{C}\tilde{x}(t), \tag{2..2}$$

where $\tilde{x} \in \mathbf{R}^{\tilde{n}}, u \in \mathbf{R}^m$, $\tilde{A}, \tilde{B}, \tilde{C}$ are real matrices of compatible dimension, and $\sigma: \mathbf{R}^m \to \mathbf{R}^m$ is a multivariable saturation nonlinearity. We assume that $\sigma(\cdot)$ is a *radial* ellipsoidal saturation function, that is, $\sigma(u)$ has the same direction as u and is confined to an ellipsoidal region in $\mathbf{R}^m$. Letting R denote an $m \times m$ positive-definite matrix, $\sigma(u)$ is defined by

$$\sigma(u) = u, \qquad\qquad u^T Ru \le 1, \tag{2..3}$$

$$= (u^T Ru)^{-\frac{1}{2}}u, \quad u^T Ru > 1. \tag{2..4}$$

Alternatively, $\sigma(u)$ can be written as

$$\sigma(u) = \beta(u)u, \tag{2..5}$$

where the function $\beta: \mathbf{R}^m \to (0, 1]$ is defined by

$$\beta(u) = 1, \qquad\qquad u^T Ru \le 1, \tag{2..6}$$

$$= \frac{1}{\sqrt{u^T Ru}}, \quad u^T Ru > 1. \tag{2..7}$$

Figure 8.1 shows the ellipsoidal function for the case $m = 2$. The closed-loop system (2..1), (2..2) can be represented by the block diagram shown in Figure 8.2. Note that in the SISO case $m = 1$, the function $u - \sigma(u)$ is a deadzone nonlinearity.

The following result provides the foundation for our synthesis approach.

Theorem 2.1.. Let $\tilde{R}_1 \in \mathbf{N}^{\tilde{n}}, R_2 \in \mathbf{P}^m, R_0 \in \mathbf{P}^m$, $\beta_0 \in [0, 1]$, and assume that $(\tilde{A}, \tilde{C})$ is observable. Furthermore, suppose there exists $\tilde{P} \in \mathbf{P}^{\tilde{n}}$ satisfying

$$\tilde{A}^T \tilde{P} + \tilde{P}\tilde{A} + \tilde{R}_1 + \tilde{C}^T R_2 \tilde{C}$$
$$+ \tfrac{1}{2}[\tilde{B}^T \tilde{P} - (1 - \beta_0)R_0\tilde{C}]^T R_0^{-1}[\tilde{B}^T \tilde{P} - (1 - \beta_0)R_0\tilde{C}] = 0. \tag{2..8}$$

Then the closed-loop system (2..1) and (2..2) is asymptotically stable with Lyapunov function $V(\tilde{x}) = \tilde{x}^T \tilde{P}\tilde{x}$, and the set

$$\tilde{\mathcal{D}} \triangleq \{\tilde{x}_0 \in \mathbf{R}^{\tilde{n}} : V(\tilde{x}_0) < \beta_\ell^{-2}\lambda_{\max}^{-1}(\tilde{C}^T R\tilde{C}\tilde{P}^{-1})\} \tag{2..9}$$

is a subset of the domain of attraction of the closed-loop system, where

$$\beta_\ell \overset{\triangle}{=} \max\left\{0, \frac{1}{2}\left[1 + \beta_0 - \sqrt{(1-\beta_0)^2 + 2\lambda_{\min}(R_2 R_0^{-1})}\right]\right\}.$$

Proof. First note that by using (2..1) and (2..2), $\dot{V}(\tilde{x}(t))$ can be written as

$$\dot{V}(\tilde{x}(t)) = -[\tilde{x}^{\mathrm{T}}(t) \ \phi^{\mathrm{T}}(u(t))]\begin{bmatrix} -\tilde{A}^{\mathrm{T}}\tilde{P} - \tilde{P}\tilde{A} & \tilde{P}\tilde{B} \\ \tilde{B}^{\mathrm{T}}\tilde{P} & 0 \end{bmatrix}\begin{bmatrix} \tilde{x}(t) \\ \phi(u(t)) \end{bmatrix},$$

where $\phi(u) \overset{\triangle}{=} u - \sigma(u)$.

Adding and subtracting $2[(1-\beta_0)u^{\mathrm{T}}(t) - \phi^{\mathrm{T}}(u(t))]R_0\phi(u(t))$ and using (2..8) yields

$$\begin{aligned}\dot{V}(\tilde{x}(t)) &= -[\tilde{x}^{\mathrm{T}}(t) \ \phi^{\mathrm{T}}(u(t))]\mathcal{M}\begin{bmatrix} \tilde{x}(t) \\ \phi(u(t)) \end{bmatrix} \\ &\quad -2(\beta(u(t)) - \beta_0)(1 - \beta(u(t)))u^{\mathrm{T}}(t)R_0 u(t) \\[2mm] &= -\tfrac{1}{2}[(\tilde{B}^{\mathrm{T}}\tilde{P} - (1-\beta_0)R_0\tilde{C})\tilde{x}(t) + 2R_0\phi(u(t))]^{\mathrm{T}}R_0^{-1}[(\tilde{B}^{\mathrm{T}}\tilde{P} \\ &\quad -(1-\beta_0)R_0\tilde{C})\tilde{x}(t) + 2R_0\phi(u(t))] \\ &\quad -\tilde{x}^{\mathrm{T}}(t)\tilde{R}_1\tilde{x}(t) - u^{\mathrm{T}}(t)[2(\beta(u(t)) - \beta_0)(1 - \beta(u(t)))R_0 + R_2]u(t). \end{aligned}$$
$$(2..10)$$

where $\mathcal{M}$ is defined by :

$$\mathcal{M} = \begin{bmatrix} \tilde{R}_1 + \tilde{C}^{\mathrm{T}}R_2\tilde{C} + \frac{1}{2}(\tilde{P}\tilde{B} - (1-\beta_0)\tilde{C}^{\mathrm{T}}R_0)R_0^{-1}(\tilde{B}^{\mathrm{T}}\tilde{P} - (1-\beta_0)R_0\tilde{C}) & \tilde{P}\tilde{B} - (1-\beta_0)\tilde{C}^{\mathrm{T}}R_0 \\ \tilde{B}^{\mathrm{T}}\tilde{P} - (1-\beta_0)R_0\tilde{C} & 2R_0 \end{bmatrix}$$

To guarantee that $\dot{V}(\tilde{x}(t)) \leq 0$, we need to show that $2(\beta(u(t)) - \beta_0)(1 - \beta(u(t)))R_0 + R_2$ is nonnegative definite for all $t \geq 0$. To do this, note that for all $t \in [0, \infty)$ it follows that

$$\begin{aligned}&2(\beta(u(t)) - \beta_0)(1 - \beta(u(t)))R_0 + R_2 \\ &= 2R_0^{\frac{1}{2}}[(\beta(u(t)) - \beta_0)(1 - \beta(u(t)))I_m + \tfrac{1}{2}R_0^{-\frac{1}{2}}R_2R_0^{-\frac{1}{2}}]R_0^{\frac{1}{2}} \\ &= 2R_0^{\frac{1}{2}}[(-\beta^2(u(t)) + \beta(u(t)) + \beta(u(t))\beta_0 - \beta_0 \\ &\quad + \tfrac{1}{2}\lambda_{\min}(R_0^{-\frac{1}{2}}R_2R_0^{-\frac{1}{2}}))I_m \\ &\quad + \tfrac{1}{2}R_0^{-\frac{1}{2}}R_2R_0^{-\frac{1}{2}} - \tfrac{1}{2}\lambda_{\min}(R_0^{-\frac{1}{2}}R_2R_0^{-\frac{1}{2}})I_m]R_0^{\frac{1}{2}}. \end{aligned}$$

If $\beta_0 \leq \frac{1}{2}\lambda_{\min}(R_0^{-\frac{1}{2}}R_2R_0^{-\frac{1}{2}}) = \frac{1}{2}\lambda_{\min}(R_2R_0^{-1})$, which is equivalent to $\beta_\ell = 0$, it is easy to check that $2(\beta(u(t)) - \beta_0)(1 - \beta(u(t)))R_0 + R_2 > 0$ for all $t \in [0, \infty)$. Thus, $\dot{V}(\tilde{x}(t)) \leq 0$ for all $t \in [0, \infty)$. If $\dot{V}(\tilde{x}(t)) = 0$, for all $t \geq 0$, it follows from (2..10) that $u(t) = \tilde{C}\tilde{x}(t) = 0$, which gives $\tilde{x}(t) = \exp(\tilde{A}t)\tilde{x}_0$, and thus $\tilde{C}\tilde{x}(t) = \tilde{C}\exp(\tilde{A}t)\tilde{x}_0 = 0$. Since $(\tilde{A}, \tilde{C})$ is observable, the invariant set consists of $\tilde{x} = 0$. It thus follows that $V(\tilde{x}(t)) \to 0$ as $t \to \infty$ and closed-loop system (2..1), (2..2) is asymptotically stable.

On the other hand, suppose that $\beta_0 > \frac{1}{2}\lambda_{\min}(R_2R_0^{-1})$. In this case $(1 - \beta_0)^2 + 2\lambda_{\min}(R_2R_0^{-1}) < (1+\beta_0)^2$ and thus

$$\beta_\ell = \frac{1}{2}[1 + \beta_0 - \sqrt{(1 - \beta_0)^2 + 2\lambda_{\min}(R_2 R_0^{-1})}].$$

Furthermore, we have the identity

$$2[\beta(u(t)) - \beta_0][1 - \beta(u(t))]R_0 + R_2$$
$$= R_0^{\frac{1}{2}}\{2[\beta(u(t)) - \beta_\ell][\frac{1}{2}[(1 + \beta_0) + \sqrt{(1 - \beta_0)^2 + 2\lambda_{\min}(R_2 R_0^{-1})}]$$
$$-\beta(u(t))]I_m$$
$$+ R_0^{-\frac{1}{2}} R_2 R_0^{-\frac{1}{2}} - \lambda_{\min}(R_0^{-\frac{1}{2}} R_2 R_0^{-\frac{1}{2}})I_m\}R_0^{\frac{1}{2}}.$$

$$(2..11)$$

Also note that for all $\beta_0 \in [0,1]$ it is easy to check that $\frac{1}{2}[(1 + \beta_0) + \sqrt{(1 - \beta_0)^2 + 2\lambda_{\min}(R_2 R_0^{-1})}] > 1$.

For all $t \in [0, \infty)$ our goal is to show that $\beta(u(t)) > \beta_\ell$, so that $2(\beta(u(t)) - \beta_0)(1 - \beta(u(t)))R_0 + R_2 > 0$, and $\dot{V}(\tilde{x}(t)) \le 0$. Let $t = 0$ and $\tilde{x}_0 \in \mathcal{D}$. If $u^T(0)Ru(0) > 1$, then, by (2..7) and (2..9),

$$\frac{1}{\beta^2(u(0))} = u^T(0)Ru(0) = \tilde{x}_0^T \tilde{C}^T R\tilde{C}\tilde{x}_0 \le \tilde{x}_0^T \tilde{P}\tilde{x}_0 \lambda_{\max}(\tilde{C}^T R\tilde{C}\tilde{P}^{-1}) < \frac{1}{\beta_\ell^2},$$

so that $\beta(u(0)) > \beta_\ell$ and hence $\dot{V}(\tilde{x}(0)) \le 0$. If, on the other hand, $u^T(0)Ru(0) \le 1$, then $\beta(u(0)) = 1$. In this case we also have $\dot{V}(\tilde{x}(0)) \le 0$. Two cases, that is, $\dot{V}(\tilde{x}(0)) < 0$ and $\dot{V}(\tilde{x}(0)) = 0$, will be treated separately.

First consider the case $\dot{V}(\tilde{x}(0)) < 0$. Suppose on the contrary there exist $T_1 > T > 0$ such that $\dot{V}(\tilde{x}(t)) < 0$ for all $t \in [0, T)$, $\dot{V}(\tilde{x}(T)) = 0$, and $\dot{V}(\tilde{x}(t)) > 0, t \in (T, T_1]$. Since $\dot{V}(\tilde{x}(t)) < 0, t \in [0, T)$, there exists T_2 satisfying $T < T_2 \le T_1$ and sufficiently close to T such that $\tilde{x}^T(t)\tilde{P}\tilde{x}(t) = V(\tilde{x}(t)) < V(\tilde{x}_0) = \tilde{x}_0^T \tilde{P}\tilde{x}_0, t \in (0, T_2]$, and thus

$$u^T(t)Ru(t) \le \tilde{x}^T(t)\tilde{P}\tilde{x}(t)\lambda_{\max}(\tilde{C}^T R\tilde{C}\tilde{P}^{-1}) < \tilde{x}_0^T \tilde{P}\tilde{x}_0 \lambda_{\max}(\tilde{C}^T R\tilde{C}\tilde{P}^{-1}) < \frac{1}{\beta_\ell^2},$$

$t \in [0, T_2]$. Hence, $\beta(u(t)) > \beta_\ell, t \in [0, T_2]$. Since $\dot{V}(\tilde{x}(t)) > 0, t \in (T, T_1]$, it follows from (2..10) and (2..11) that $\beta(u(t)) < \beta_\ell, t \in (T, T_1]$. Therefore, $\beta(u(T_2)) < \beta_\ell$, which is a contradiction. As a result, $\dot{V}(\tilde{x}(t)) \le 0$ for all $t \ge 0$. Again, using the assumption that $(\tilde{A}, \tilde{C})$ is observable, we conclude that the closed-loop system (2..1), (2..2) is asymptotically stable.

Next, consider the case $\dot{V}(\tilde{x}(0)) = 0$. It follows from (2..10) that $u(0) = 0$, that is, $u^T(0)Ru(0) = 0$. Furthermore, for $t > 0$, $\dot{V}(\tilde{x}(t)) > 0$ implies that $\beta(u(t)) < 1$, that is, $u^T(t)Ru(t) > 1$. For t sufficiently close to 0, however, this condition violates the continuity of $u(t)$. It thus follows that there exists $T_0 > 0$ sufficiently close to 0 such that $\dot{V}(\tilde{x}(t)) \le 0$ for all $t \in (0, T_0]$. Using similar arguments as in the case $\dot{V}(\tilde{x}(0)) < 0$, it can be shown that $\dot{V}(\tilde{x}(t)) \ne 0$ for all $t \in (0, T_0]$. Therefore, $\dot{V}(\tilde{x}(t)) < 0$ for all $t \in (0, T_0]$. In particular, $\dot{V}(\tilde{x}(T_0)) < 0$. Hence we can proceed as in the previous case where $\dot{V}(\tilde{x}(0)) < 0$ with the time 0 replaced by T_0. It thus follows that $\dot{V}(\tilde{x}(t)) \to 0$ as $t \to \infty$ and the closed-loop system (2..1), (2..2) is asymptotically stable.

Remark 2.1.. From the proof of Theorem 2.1, it is easy to see that β_l indeed is a lower bound of $\beta(u(t))$ for all $t \geq 0$. Note that if $\beta_\ell = 0$, then $\tilde{\mathcal{D}} = \mathbf{R}^{\tilde{n}}$.

Remark 2.2.. Theorem 2.1 can be viewed as an application of the positive real lemma [18] to a deadzone nonlinearity. To see this, define

$$\tilde{L}^T \triangleq [-(\tilde{B}^T \tilde{P} - R_0 \tilde{C}(1 - \beta_0))^T (2R_0)^{-\frac{1}{2}} \quad (\tilde{R}_1 + \tilde{C}^T R_2 \tilde{C})^{\frac{1}{2}}] V,$$
$$\tilde{W}^T \triangleq [(2R_0)^{\frac{1}{2}} \quad 0] V,$$

where $V^T V = I$. It is easy to check that the equations

$$0 = \tilde{A}^T \tilde{P} + \tilde{P}\tilde{A} + \tilde{L}^T \tilde{L},$$
$$0 = \tilde{P}\tilde{B} - (1 - \beta_0)\tilde{C}^T R_0 + \tilde{L}^T \tilde{W},$$
$$0 = 2R_0 - \tilde{W}^T \tilde{W},$$

are satisfied and are equivalent to the Riccati equation (2..8). It thus follows that $\tilde{G}(s)$ is positive real, where $\tilde{G}(s) \sim \left[\begin{array}{c|c} \tilde{A} & \tilde{B} \\ \hline (1 - \beta_0)R_0\tilde{C} & R_0 \end{array} \right]$.

Remark 2.3.. The small gain theorem can be viewed as a special case of Theorem 2.1. This can be verified by using a simple loopshifting technique. First, note that the closed-loop (2..1), (2..2) can be written as

$$\dot{\tilde{x}}(t) = (\tilde{A} - \tfrac{1}{2}\tilde{B}\tilde{C})\tilde{x}(t) + \tilde{B}(\sigma(u(t)) - \tfrac{1}{2}u(t)), \quad \tilde{x}(0) = \tilde{x}_0, \quad (2..12)$$
$$u(t) = \tilde{C}\tilde{x}(t), \quad\quad\quad\quad\quad\quad\quad\quad\quad\quad\quad\quad\quad\quad\quad\quad\quad (2..13)$$

and it is easy to check that the nonlinearity $\sigma(u(t)) - \tfrac{1}{2}u(t)$ is bounded by the sector $[-\tfrac{1}{2}I, \tfrac{1}{2}I]$. Next, by choosing $\beta_0 = 0$, $R_0 = 2I$, $\tilde{R}_1 = 0$, $R_2 = 0$, equation (2..8) can be reduced to the Riccati equation

$$0 = (\tilde{A} - \tfrac{1}{2}\tilde{B}\tilde{C})^T \tilde{P} + \tilde{P}(\tilde{A} - \tfrac{1}{2}\tilde{B}\tilde{C}) + \tilde{C}^T\tilde{C} + \tfrac{1}{4}\tilde{P}\tilde{B}\tilde{B}^T\tilde{P}, \quad\quad (2..14)$$

which implies that

$$\left\| \left[\begin{array}{c|c} \tilde{A} - \tfrac{1}{2}\tilde{B}\tilde{C} & \tilde{B} \\ \hline \tilde{C} & 0 \end{array} \right] \right\|_\infty \leq 2. \quad\quad (2..15)$$

Proposition 2.1. Suppose that the assumptions of Theorem 2.1 are satisfied and $\beta_0 \leq \tfrac{1}{2}\lambda_{\min}(R_2 R_0^{-1})$. Then the closed-loop system (2..1), (2..2) is globally asymptotically stable. Furthermore, all the eigenvalues of $\tilde{A} - \tilde{B}\tilde{C}$ lie in the closed left half plane.

Proof. If $\beta_0 \leq \tfrac{1}{2}\lambda_{\min}(R_2 R_0^{-1})$, then $\beta_\ell = 0$. It thus follows directly from Theorem 2.1 that the closed-loop system (2..1), (2..2) is globally asymptotically stable. To show that every eigenvalue of $\tilde{A} - \tilde{B}\tilde{C}$ has nonpositive real part, rewrite (2..8) as

$$\begin{aligned} 0 = & (\tilde{A} - \tilde{B}\tilde{C})^T \tilde{P} + \tilde{P}(\tilde{A} - \tilde{B}\tilde{C}) + [\tilde{B}^T \tilde{P} + (1 + \beta_0)R_0\tilde{C}]^T (2R_0)^{-1}[\tilde{B}^T \tilde{P} \\ & + (1 + \beta_0)R_0\tilde{C}] + \tilde{R}_1 + \tilde{C}^T (R_2 - 2\beta_0 R_0)\tilde{C}, \end{aligned}$$
$$\quad\quad (2..16)$$

and note that $\beta_0 \leq \frac{1}{2}\lambda_{\min}(R_2 R_0^{-1})$ implies that $R_2 - 2\beta_0 R_0 \geq 0$. Next, let $\lambda \in \mathbf{C}$ and $z \in \mathbf{C}^{\tilde{n}}, z \neq 0$, satisfy $(\tilde{A} - \tilde{B}\tilde{C})z = \lambda z$. Forming $z^* (2..16) z$ yields

$$
\begin{aligned}
0 = \; & 2z^* \tilde{P} z \, \mathrm{Re}[\lambda] + z^*[(\tilde{B}^T \tilde{P} + (1+\beta_0)R_0 \tilde{C})^T (2R_0)^{-1}(\tilde{B}^T \tilde{P} + (1+\beta_0)R_0 \tilde{C}) \\
& + \tilde{R}_1 + \tilde{C}^T (R_2 - 2\beta_0 R_0)\tilde{C}]z.
\end{aligned}
$$

Since $\tilde{P} > 0$ and $(\tilde{B}^T \tilde{P} + (1+\beta_0)R_0 \tilde{C})^T (2R_0)^{-1}(\tilde{B}^T \tilde{P} + (1+\beta_0)R_0 \tilde{C}) + \tilde{R}_1 + \tilde{C}^T (R_2 - 2\beta_0 R_0)\tilde{C} \geq 0$, it follows that $\mathrm{Re}[\lambda] \leq 0$.

The significance of Proposition 2.1 is as follows. For the linear dynamic compensator given by (3..3) and (3..4), $\tilde{A} - \tilde{B}\tilde{C} = \begin{bmatrix} A & 0 \\ B_c C & A_c \end{bmatrix}$. Similarly, for the nonlinear dynamic compensator given by (4..1) and (4..2), $\tilde{A} - \tilde{B}\tilde{C} = \begin{bmatrix} A & 0 \\ B_c C & A_c - E_c C_c \end{bmatrix}$. Hence, when $\beta_0 \leq \frac{1}{2}\lambda_{\min}(R_2 R_0^{-1})$ and Theorem 2.1 guarantees global stability of the closed-loop system subject to input saturation, the poles of the open-loop system must lie in the closed left half plane, which is consistent with the well-known result [8].

3. Linear Controller Synthesis

In this section, we consider linear controller synthesis based upon Theorem 2.1. Consider the plant

$$
\begin{aligned}
\dot{x}(t) &= Ax(t) + B\sigma(u(t)), \quad x(0) = x_0, & (3..1) \\
y(t) &= Cx(t), & (3..2)
\end{aligned}
$$

where $x \in \mathbf{R}^n, u \in \mathbf{R}^m, y \in \mathbf{R}^l$, (A, B) is controllable, (A, C) is observable, and let the dynamic compensator have the form

$$
\begin{aligned}
\dot{x}_c(t) &= A_c x_c(t) + B_c y(t), \quad x_c(0) = x_{c0}, & (3..3) \\
u(t) &= C_c x_c(t), & (3..4)
\end{aligned}
$$

where $x_c \in \mathbf{R}^{n_c}$ and $n_c \leq n$. Then the closed-loop system can be written in the form of (2..1), (2..2) with

$$
\tilde{x} \triangleq \begin{bmatrix} x \\ x_c \end{bmatrix}, \quad \tilde{x}_0 \triangleq \begin{bmatrix} x_0 \\ x_{c0} \end{bmatrix},
$$

$$
\tilde{A} \triangleq \begin{bmatrix} A & BC_c \\ B_c C & A_c \end{bmatrix}, \quad \tilde{B} \triangleq \begin{bmatrix} B \\ 0 \end{bmatrix}, \quad \tilde{C} \triangleq [\, 0 \quad C_c \,].
$$

Our goal is to determine gains A_c, B_c, C_c that minimize the LQG-type cost

$$
J(A_c, B_c, C_c) = \mathrm{tr}\, \tilde{P}\tilde{V}, \qquad (3..5)
$$

where $\tilde{V} = \begin{bmatrix} V_1 & 0 \\ 0 & B_c V_2 B_c^T \end{bmatrix}$, $\tilde{P}$ satisfies (2..8), and $V_1 \in \mathbf{N}^n$ and $V_2 \in \mathbf{P}^l$ are analogous to the plant disturbance and measurement noise intensity matrices of LQG theory, respectively. Furthermore, let $\tilde{R}_1 = \begin{bmatrix} R_1 & 0 \\ 0 & 0 \end{bmatrix}$, where $R_1 \in \mathbf{N}^n$.

We first consider the reduced-order controller case, that is, $n_c \leq n$. The following results are obtained by minimizing $J(A_c, B_c, C_c)$ with respect to A_c, B_c, C_c. These necessary conditions then provide sufficient conditions for closed-loop stability by applying Theorem 2.1. For convenience define $\Sigma_1 \triangleq \frac{1}{4}(1 + \beta_0)^2 B(R_2 + \frac{1}{2}(1 - \beta_0)^2 R_0)^{-1} B^T$, $\Sigma_0 \triangleq \frac{1}{2} B R_0^{-1} B^T$ and $\overline{\Sigma} \triangleq C^T V_2^{-1} C$.

Before we consider the reduced-order case, the following lemma is required.

Lemma 3.1. [17] Let $\hat{P}, \hat{Q}$ be $n \times n$ nonnegative-definite matrices and suppose that rank $\hat{Q}\hat{P} = n_c$. Then there exist $n_c \times \tilde{n}$ matrices G, Γ and an $n_c \times n_c$ invertible matrix M, unique except for a change of basis in $\mathbf{R}^{n_c}$, such that

$$\hat{Q}\hat{P} = G^T M \Gamma, \quad \Gamma G^T = I_{n_c}. \tag{3..6}$$

Furthermore, the $n \times n$ matrices

$$\tau \triangleq G^T \Gamma, \quad \tau_\perp \triangleq I_n - \tau, \tag{3..7}$$

are idempotent and have rank n_c and $n - n_c$, respectively. If, in addition, rank $\hat{Q} = $ rank $\hat{P} = n_c$, then

$$\tau\hat{Q} = \hat{Q}, \quad \hat{P}\tau = \hat{P}. \tag{3..8}$$

Proposition 3.1. Let $n_c \leq n, \beta_0 \in [0, 1]$, suppose there exist $n \times n$ nonnegative-definite matrices $P, Q, \hat{P}, \hat{Q}$ satisfying

$$0 = A^T P + PA + R_1 - P(\Sigma_1 - \Sigma_0)P + \tau_\perp^T P \Sigma_1 P \tau_\perp, \tag{3..9}$$

$$0 = (A - Q\overline{\Sigma} + \Sigma_0 P)^T \hat{P} + \hat{P}(A - Q\overline{\Sigma} + \Sigma_0 P) + \hat{P}\Sigma_0\hat{P}$$
$$+ P\Sigma_1 P - \tau_\perp^T P \Sigma_1 P \tau_\perp, \tag{3..10}$$

$$0 = [A + \Sigma_0(P + \hat{P})]Q + Q[A + \Sigma_0(P + \hat{P})]^T + V_1$$
$$- Q\overline{\Sigma}Q + \tau_\perp Q\overline{\Sigma}Q\tau_\perp^T, \tag{3..11}$$

$$0 = [A + (\Sigma_0 - \Sigma_1)P]\hat{Q} + \hat{Q}[A + (\Sigma_0 - \Sigma_1)P]^T$$
$$+ Q\overline{\Sigma}Q - \tau_\perp Q\overline{\Sigma}Q\tau_\perp^T, \tag{3..12}$$

$$\text{rank } \hat{Q} = \text{rank } \hat{P} = \text{rank } \hat{Q}\hat{P} = n_c, \tag{3..13}$$

and let A_c, B_c, C_c be given by

$$A_c = \Gamma A G^T + \frac{1}{2}(1 + \beta_0)\Gamma B C_c - B_c C G^T + \Gamma \Sigma_0 P G^T, \tag{3..14}$$

$$B_c = \Gamma Q C^T V_2^{-1}, \tag{3..15}$$

$$C_c = -\frac{1}{2}(1 + \beta_0)[R_2 + \frac{1}{2}(1 - \beta_0)^2 R_0]^{-1} B^T P G^T. \tag{3..16}$$

Furthermore, suppose that $(\tilde{A}, \tilde{C})$ is observable. Then $\tilde{P} = \begin{bmatrix} P + \hat{P} & -\hat{P}G^{\mathrm{T}} \\ -G\hat{P} & G\hat{P}G^{\mathrm{T}} \end{bmatrix}$ satisfies (2..8). Furthermore, the equilibrium solution $\tilde{x}(t) \equiv 0$ of the closed-loop system (2..1), (2..2) is asymptotically stable, and $\hat{\mathcal{D}}$ defined by (2..9) is a subset of the domain of attraction of the closed-loop system.

Proof. The result is obtained by applying the Lagrange multiplier technique to (3..5) subject to (2..8) and by partitioning $\tilde{P}$ and $\tilde{Q}$ as

$$\tilde{P} = \begin{bmatrix} P_1 & P_{12} \\ P_{12}^{\mathrm{T}} & P_2 \end{bmatrix}, \quad \tilde{Q} = \begin{bmatrix} Q_1 & Q_{12} \\ Q_{21}^{\mathrm{T}} & Q_2 \end{bmatrix}.$$

Here, we show only the key steps. First, define the Lagrangian

$$\mathcal{L} = \operatorname{tr} \tilde{P}\tilde{V} + \operatorname{tr} \tilde{Q}[(\tilde{A} - \tfrac{1}{2}(1 - \beta_0)\tilde{B}\tilde{C})^{\mathrm{T}}\tilde{P} + \tilde{P}(\tilde{A} - \tfrac{1}{2}(1 - \beta_0)\tilde{B}\tilde{C}) + \tilde{R}_1$$
$$+ \tilde{C}^{\mathrm{T}}(R_2 + \tfrac{1}{2}(1 - \beta_0)^2 R_0)\tilde{C} + \tfrac{1}{2}\tilde{P}\tilde{B}R_0^{-1}\tilde{B}^{\mathrm{T}}\tilde{P}].$$

Taking derivatives with respect to A_c, B_c, C_c and $\tilde{P}$, and setting them to zero yields

$$0 = \frac{\partial \mathcal{L}}{\partial A_c} = 2(P_{12}^{\mathrm{T}}Q_{12} + P_2 Q_2), \tag{3..17}$$

$$0 = \frac{\partial \mathcal{L}}{\partial B_c} = 2P_2 B_c V_2 + 2(P_{12}^{\mathrm{T}}Q_1 + P_2 Q_{12}^{\mathrm{T}})C^{\mathrm{T}}, \tag{3..18}$$

$$0 = \frac{\partial \mathcal{L}}{\partial C_c} = 2(R_2 + \tfrac{1}{2}(1 - \beta_0)^2 R_0)C_c Q_2 + (1 + \beta_0)B^{\mathrm{T}}(P_1 Q_{12}$$
$$+ P_{12}Q_2), \tag{3..19}$$

$$0 = \frac{\partial \mathcal{L}}{\partial \tilde{P}} = (\tilde{A} - \tfrac{1}{2}(1 - \beta_0)\tilde{B}\tilde{C} + \tfrac{1}{2}\tilde{B}R_0^{-1}\tilde{B}^{\mathrm{T}}\tilde{P})\tilde{Q}$$
$$+ \tilde{Q}(\tilde{A} - \tfrac{1}{2}(1 - \beta_0)\tilde{B}\tilde{C} + \tfrac{1}{2}\tilde{B}R_0^{-1}\tilde{B}^{\mathrm{T}}\tilde{P})^{\mathrm{T}} + \tilde{V}. \tag{3..20}$$

Next, define $P, Q, \hat{P}, \hat{Q}, \Gamma, G, M$ by

$$P \triangleq P_1 - \hat{P}, \quad \hat{P} \triangleq P_{12}P_2^{-1}P_{12}^{\mathrm{T}}, \quad Q \triangleq Q_1 - \hat{Q}, \quad \hat{Q} \triangleq Q_{12}Q_2^{-1}Q_{12}^{\mathrm{T}},$$

$$G^{\mathrm{T}} \triangleq Q_{12}Q_2^{-1}, \quad M \triangleq Q_2 P_2, \quad \Gamma \triangleq -P_2^{-1}P_{12}^{\mathrm{T}}.$$

Algebraic manipulation of equations (3..18) and (3..19) yields B_c and C_c given by (3..15) and (3..16). The expression (3..14) for A_c is obtained by combining the $(1, 2)$ and $(2, 2)$ blocks of equation (2..8) or (3..20) using (3..17). Equations (3..9) and (3..10) are obtained by combining the $(1, 1)$ and $(1, 2)$ blocks of equation (2..8). Similarly, (3..11) and (3..12) are obtained by combining the $(1, 1)$ and $(1, 2)$ blocks of equation (3..20). See [17] for details.

Remark 3.1.. Suppose $x_{c0} = 0$ and consider initial conditions of the form $\tilde{x}_0 = [x_0^T \ 0]^T$. Then, since $P_1 = P + \hat{P}$, the set $\mathcal{D} \times \{0\}$, where $\mathcal{D}$ is defined by

$$\mathcal{D} \triangleq \{x_0 \in \mathbf{R}^n : x_0^T(P + \hat{P})x_0 < \beta_\ell^{-2}\lambda_{\max}^{-1}(\tilde{C}^T R\tilde{C}\tilde{P}^{-1})\}, \qquad (3..21)$$

is a subset of $\tilde{\mathcal{D}}$, and thus is also a subset of domain of attraction.

Next, we consider full-order controller case $n_c = n$.

Proposition 3.2.. Let $n_c = n, \beta_0 \in [0,1]$, suppose there exist $n \times n$ nonnegative-definite matrices $P, Q, \hat{P}$ satisfying

$$0 = A^T P + PA + R_1 - P(\Sigma_1 - \Sigma_0)P, \qquad (3..22)$$
$$0 = (A - Q\overline{\Sigma} + \Sigma_0 P)^T\hat{P} + \hat{P}(A - Q\overline{\Sigma} + \Sigma_0 P) + \hat{P}\Sigma_0\hat{P} + P\Sigma_1 P, \qquad (3..23)$$
$$0 = [A + \Sigma_0(P + \hat{P})]Q + Q[A + \Sigma_0(P + \hat{P})]^T + V_1 - Q\overline{\Sigma}Q, \qquad (3..24)$$

and let A_c, B_c, C_c be given by

$$A_c = A + \tfrac{1}{2}(1 + \beta_0)BC_c - B_cC + \Sigma_0 P, \qquad (3..25)$$
$$B_c = QC^T V_2^{-1}, \qquad (3..26)$$
$$C_c = -\tfrac{1}{2}(1 + \beta_0)[R_2 + \tfrac{1}{2}(1 - \beta_0)^2 R_0]^{-1}B^T P. \qquad (3..27)$$

Furthermore, suppose that $(\tilde{A}, \tilde{C})$ is observable. Then $\tilde{P} = \begin{bmatrix} P + \hat{P} & -\hat{P} \\ -\hat{P} & \hat{P} \end{bmatrix}$ satisfies (2..8). Furthermore, the equilibrium solution $\tilde{x}(t) \equiv 0$ of the closed-loop system (2..1), (2..2) is asymptotically stable, and $\tilde{\mathcal{D}}$ defined by (2..9) is a subset of the domain of attraction of the closed-loop system.

Proof. The proof is a special case of the proof of Proposition 3.1 below with $n_c = n$ and $\Gamma = G^T = \tau = I$.

4. Nonlinear Controller Synthesis I

In this section we consider the plant (3..1) and (3..2) with nonlinear controllers of the form

$$\dot{x}_c(t) = A_c x_c(t) + B_c y(t) + E_c[\sigma(u(t)) - u(t)], \quad x_c(0) = x_{c0}, \quad (4..1)$$
$$u(t) = C_c x_c(t), \qquad (4..2)$$

where $x_c \in \mathbf{R}^{n_c}$ and $n_c \leq n$. This compensator includes the nonlinear term $E_c(\sigma(u(t)) - u(t))$ as in the observer-based antiwindup setup studied in [14].

With (4..1), (4..2) the closed-loop system can be written in the form of (2..1), (2..2) with

$$\tilde{x} \triangleq \begin{bmatrix} x \\ x_c \end{bmatrix}, \quad \tilde{x}_0 \triangleq \begin{bmatrix} x_0 \\ x_{c0} \end{bmatrix},$$

$$\tilde{A} \triangleq \begin{bmatrix} A & BC_c \\ B_c C & A_c \end{bmatrix}, \quad \tilde{B} \triangleq \begin{bmatrix} B \\ E_c \end{bmatrix}, \quad \tilde{C} \triangleq [\, 0 \quad C_c \,].$$

For simplicity, let $R_0 = \frac{\gamma}{1-\beta_0} R_2$, where $\gamma > 0$. Then equation (2..8) can be rewritten as

$$\tilde{A}^T \tilde{P} + \tilde{P}\tilde{A} + \tilde{R}_1 + \tilde{C}^T R_2 \tilde{C}$$
$$+ \tfrac{1}{2}\gamma^{-1}(1-\beta_0)(\tilde{B}^T\tilde{P} - \gamma R_2 \tilde{C})^T R_2^{-1}(\tilde{B}^T\tilde{P} - \gamma R_2 \tilde{C}) \quad (4..3)$$
$$= 0.$$

As before, we minimize the cost functional (3..5) subject to (4..3). The following result as well as Proposition 4.1 and later results are obtained by minimizing $J(A_c, B_c, C_c)$ with respect to A_c, B_c, C_c, with E_c chosen to have a specific form. These necessary conditions then provide sufficient conditions for closed-loop stability by applying Theorem 2.1. In the previous section equation (3..20) for $\tilde{Q}$ was obtained by differentiating the Lagrangian with respect to $\tilde{P}$. However, due to the presence of the nonlinear term involving E_c, the minimization yields inconsistent expressions for A_c as obtained from equations (2..8) and (3..20) for $\tilde{P}$ and $\tilde{Q}$. To circumvent this problem we take a suboptimal approach involving a specific choice of E_c and an alternative equation for $\tilde{Q}$ which yields consistent expressions for A_c. For convenience define $\Sigma \triangleq BR_2^{-1}B^T$ and $\delta \triangleq 2 + \gamma(1 - \beta_0)$.

Proposition 4.1. Let $n_c \leq n, \gamma > 0, \beta_0 \in [0,1]$, suppose there exist $n \times n$ nonnegative-definite matrices $P, Q, \hat{P}, \hat{Q}$ satisfying (3..13) and

$$0 = A^T P + PA + R_1 - \tfrac{1}{2}[\delta^{-1}(1+\beta_0)^2 - \gamma^{-1}(1-\beta_0)]P\Sigma P$$
$$+ \tfrac{1}{2}\delta^{-1}(1+\beta_0)^2 \tau_\perp^T P\Sigma P\tau_\perp, \quad (4..4)$$

$$0 = [A - Q\overline{\Sigma} + \tfrac{1}{4}\gamma^{-1}(1-\beta_0)^2\Sigma P]^T\hat{P} + \hat{P}[A - Q\overline{\Sigma} + \tfrac{1}{4}\gamma^{-1}(1-\beta_0)^2\Sigma P]$$
$$+ \tfrac{1}{8}\gamma^{-1}(1-\beta_0)^3\hat{P}\Sigma\hat{P} + \tfrac{1}{2}\delta^{-1}(1+\beta_0)^2(P\Sigma P - \tau_\perp^T P\Sigma P\tau_\perp), \quad (4..5)$$

$$0 = AQ + QA^T + V_1 - Q\overline{\Sigma}Q + \tau_\perp Q\overline{\Sigma}Q\tau_\perp^T, \quad (4..6)$$

$$0 = [A - \tfrac{1}{2}\delta^{-1}(1+\beta_0)^2\Sigma P]\hat{Q} + \hat{Q}[A - \tfrac{1}{2}\delta^{-1}(1+\beta_0)^2\Sigma P]^T$$
$$+ Q\overline{\Sigma}Q - \tau_\perp Q\overline{\Sigma}Q\tau_\perp^T, \quad (4..7)$$

and let A_c, B_c, C_c, E_c be given by

$$A_c = \Gamma AG^T + \tfrac{1}{4}(1+\beta_0)(3-\beta_0)\Gamma BC_c - B_c CG^T$$
$$+ \tfrac{1}{4}\gamma^{-1}(1-\beta_0)^2\Gamma\Sigma PG^T, \quad (4..8)$$

$$B_c = \Gamma QC^TV_2^{-1}, \quad (4..9)$$

$$C_c = -\delta^{-1}(1+\beta_0)R_2^{-1}B^TPG^T, \quad (4..10)$$

$$E_c = \tfrac{1}{2}(1+\beta_0)\Gamma B. \quad (4..11)$$

Furthermore, suppose that $(\tilde{A}, \tilde{C})$ is observable. Then $\tilde{P} = \begin{bmatrix} P + \hat{P} & -\hat{P}G^{\mathrm{T}} \\ -G\hat{P} & G\hat{P}G^{\mathrm{T}} \end{bmatrix}$
satisfies (2..8). Furthermore, the equilibrium solution $\tilde{x}(t) \equiv 0$ of the closed-loop system (2..1), (2..2) is asymptotically stable, and $\tilde{\mathcal{D}}$ defined by (2..9) is a subset of the domain of attraction of the closed-loop system.

Proof. Let $E_c = \frac{1}{2}(1 + \beta_0)\Gamma B$ and require that $\tilde{Q}$ satisfy

$$0 = (\tilde{A} + \begin{bmatrix} -\frac{1}{2}(1 - \beta_0)B \\ \kappa E_c \end{bmatrix} \tilde{C})\tilde{Q} + \tilde{Q}(\tilde{A} + \begin{bmatrix} -\frac{1}{2}(1 - \beta_0)B \\ \kappa E_c \end{bmatrix} \tilde{C})^{\mathrm{T}} + \tilde{V}, \quad (4..12)$$

where $\kappa \triangleq \frac{1 - \beta_0}{2} - \frac{(1 - \beta_0)^2 \delta}{2\gamma(1 + \beta_0)^2}$. The remaining steps are similar to the proof of Proposition 5.1.

Next we consider the full order case $n_c = n$.

Proposition 4.2. Let $n_c = n, \gamma > 0$, and $\beta_0 \in [0, 1]$, suppose there exist $n \times n$ nonnegative-definite matrices $P, Q, \hat{P}$ satisfying

$$0 = A^{\mathrm{T}}P + PA + R_1 - \frac{1}{2}[\delta^{-1}(1 + \beta_0)^2 - \gamma^{-1}(1 - \beta_0)]P\Sigma P, \quad (4..13)$$

$$0 = [A - Q\overline{\Sigma} + \frac{1}{4}\gamma^{-1}(1 - \beta_0)^2\Sigma P]^{\mathrm{T}}\hat{P} + \hat{P}[A - Q\overline{\Sigma} + \frac{1}{4}\gamma^{-1}(1 - \beta_0)^2\Sigma P]$$
$$+ \frac{1}{8}\gamma^{-1}(1 - \beta_0)^3\hat{P}\Sigma\hat{P} + \frac{1}{2}\delta^{-1}(1 + \beta_0)^2 P\Sigma P, \quad (4..14)$$

$$0 = AQ + QA^{\mathrm{T}} + V_1 - Q\overline{\Sigma}Q, \quad (4..15)$$

and let A_c, B_c, C_c, E_c be given by

$$A_c = A + \frac{1}{4}(1 + \beta_0)(3 - \beta_0)BC_c - B_cC$$
$$+ \frac{1}{4}\gamma^{-1}(1 - \beta_0)^2\Sigma P, \quad (4..16)$$

$$B_c = QC^{\mathrm{T}}V_2^{-1}, \quad (4..17)$$

$$C_c = -\delta^{-1}(1 + \beta_0)R_2^{-1}B^{\mathrm{T}}P, \quad (4..18)$$

$$E_c = \frac{1}{2}(1 + \beta_0)B. \quad (4..19)$$

Furthermore, suppose that $(\tilde{A}, \tilde{C})$ is observable. Then $\tilde{P} = \begin{bmatrix} P + \hat{P} & -\hat{P} \\ -\hat{P} & \hat{P} \end{bmatrix}$
satisfies (2..8). Furthermore, the equilibrium solution $\tilde{x}(t) \equiv 0$ of the closed-loop system (2..1), (2..2) is asymptotically stable, and $\tilde{\mathcal{D}}$ defined by (2..9) is a subset of the domain of attraction of the closed-loop system.

Proof. The result is a special case of Proposition 4.1 with $n_c = n$ and $\Gamma = G^{\mathrm{T}} = \tau = I$.

Remark 4.1. Note that in Proposition 4.2, equations (4..13)-(4..15) are coupled in one direction, so that no iteration is required to solve them.

In the full-order case with $\beta_0 = 1$, the dynamic compensator given by Proposition 4.2 is an observer based controller with the realization

$$\dot{x}_c(t) = Ax_c(t) + B\sigma(u(t)) + B_c(y(t) - Cx_c(t)),$$
$$u(t) = C_c x_c(t),$$

where B_c and C_c are the standard LQG estimator and controller gains, respectively. Furthermore, since $\beta_0 = 1$, it can be seen from the proof of Theorem 2.1 that $\beta(u(t)) = 1$ for all $t \geq 0$. Hence the guaranteed domain of attraction of the closed-loop system yields only unsaturated control signals, that is, $\sigma(u(t)) = u(t)$ for all $t \geq 0$.

5. Nonlinear Controller Synthesis II

In this section we consider again the nonlinear controller given by (4..1), (4..2), and develop an alternative approach for obtaining controller gains. Specifically, in place of (4..12) we require that $\tilde{Q}$ satisfy the alternative equation (5..8).

Proposition 5.1. Let $n_c \leq n, \gamma > 0, \beta_0 \in [0,1]$, suppose there exist $n \times n$ nonnegative-definite matrices $P, Q, \hat{P}, \hat{Q}$ satisfying (3..13) and

$$
\begin{aligned}
0 = \ & A^T P + PA + R_1 - \tfrac{1}{2}[\delta^{-1}(1+\beta_0)^2 - \gamma^{-1}(1-\beta_0)]P\Sigma P \\
& + \tfrac{1}{2}\delta^{-1}(1+\beta_0)^2 \tau_\perp^T P\Sigma P\tau_\perp,
\end{aligned}
\tag{5..1}
$$

$$
\begin{aligned}
0 = \ & [A - Q\overline{\Sigma} + \tfrac{1}{4}\gamma^{-1}(1-\beta_0)^2\Sigma P]^T \hat{P} + \hat{P}[A - Q\overline{\Sigma} + \tfrac{1}{4}\gamma^{-1}(1-\beta_0)^2\Sigma P] \\
& + \tfrac{1}{8}\gamma^{-1}(1-\beta_0)^3 \hat{P}\Sigma\hat{P} + \tfrac{1}{2}\delta^{-1}(1+\beta_0)^2(P\Sigma P - \tau_\perp^T P\Sigma P\tau_\perp),
\end{aligned}
\tag{5..2}
$$

$$
\begin{aligned}
0 = \ & [A + \tfrac{1}{4}(\gamma^{-1}(1-\beta_0)^2 - \delta^{-1}(1-\beta_0)(1+\beta_0)^2)\Sigma(P+\hat{P})]Q \\
& + Q[A + \tfrac{1}{4}(\gamma^{-1}(1-\beta_0)^2 - \delta^{-1}(1-\beta_0)(1+\beta_0)^2)\Sigma(P+\hat{P})]^T \\
& + V_1 - Q\overline{\Sigma}Q + \tau_\perp Q\overline{\Sigma}Q\tau_\perp^T,
\end{aligned}
\tag{5..3}
$$

$$
\begin{aligned}
0 = \ & [A + \tfrac{1}{4}(\gamma^{-1}(1-\beta_0)^2 - \delta^{-1}(3-\beta_0)(1+\beta_0)^2)\Sigma P]\hat{Q} \\
& + \hat{Q}[A + \tfrac{1}{4}(\gamma^{-1}(1-\beta_0)^2 - \delta^{-1}(3-\beta_0)(1+\beta_0)^2)\Sigma P]^T \\
& + Q\overline{\Sigma}Q - \tau_\perp Q\overline{\Sigma}Q\tau_\perp^T,
\end{aligned}
\tag{5..4}
$$

and let A_c, B_c, C_c, E_c be given by (4..8)-(4..11). Furthermore, suppose that $(\tilde{A}, \tilde{C})$ is observable. Then $\tilde{P} = \begin{bmatrix} P + \hat{P} & -\hat{P}G^T \\ -G\hat{P} & G\hat{P}G^T \end{bmatrix}$ satisfies (2..8). Furthermore, the equilibrium solution $\tilde{x}(t) \equiv 0$ of the closed-loop system (2..1), (2..2) is asymptotically stable, and $\tilde{D}$ defined by (2..9) is a subset of the domain of attraction of the closed-loop system.

Proof. Defining the Lagrangian

$$
\begin{aligned}
\mathcal{L} = \ & \operatorname{tr} \tilde{P}\tilde{V} + \operatorname{tr} \tilde{Q}[\tilde{A}^T\tilde{P} + \tilde{P}\tilde{A} + \tilde{R}_1 + \tilde{C}^T R_2 \tilde{C} \\
& + \tfrac{1-\beta_0}{2\gamma}(\tilde{B}^T\tilde{P} - \gamma R_2\tilde{C})^T R_2^{-1}(\tilde{B}^T\tilde{P} - \gamma R_2\tilde{C})]
\end{aligned}
$$

yields

$$0 = \frac{\partial \mathcal{L}}{\partial A_c} = 2(P_{12}^T Q_{12} + P_2 Q_2), \tag{5..5}$$

$$0 = \frac{\partial \mathcal{L}}{\partial B_c} = 2P_2 B_c V_2 + 2(P_{12}^T Q_1, P_2 Q_{12}^T) C^T, \tag{5..6}$$

$$0 = \frac{\partial \mathcal{L}}{\partial C_c} = [1 + \tfrac{1}{2}\gamma(1 - \beta_0)] R_2 C_c Q_2 + [1 - \tfrac{1}{2}(1 - \beta_0)] B^T (P_1 Q_{12}$$
$$+ P_{12} Q_2). \tag{5..7}$$

Next, let $E_c = \tfrac{1}{2}(1 + \beta_0)\Gamma B$ and require that $\tilde{Q}$ satisfy

$$0 = \tilde{V}$$
$$+ [\tilde{A} - \tfrac{1}{2}(1 - \beta_0)\tilde{B}_0 \tilde{C} + k \tilde{B}_0 R_2^{-1} \tilde{B}_0^T \tilde{P}] \tilde{Q} + \tilde{Q} [\tilde{A} - \tfrac{1}{2}(1 - \beta_0)\tilde{B}_0 \tilde{C} + k \tilde{B}_0 R_2^{-1} \tilde{B}_0^T \tilde{P}]^T, \tag{5..8}$$

where $k \triangleq \tfrac{1}{4}\gamma^{-1}(1 - \beta_0)^2 - \tfrac{1}{4}\delta^{-1}(1 - \beta_0)(1 + \beta_0)^2$ and $\tilde{B}_0 \triangleq \begin{bmatrix} B \\ 0 \end{bmatrix}$. The
remaining steps are similar to the proof of Proposition 3.1.

Note that since $\tilde{Q}$ satisfies the alternative equation (5..8) in place of
(4..12), the synthesis equations (4..13)-(4..15) and (4..4)-(4..7) are different
from (5..9)-(5..11) and (5..1)-(5..4). However, the expressions for A_c, B_c, C_c
remain unchanged.

Proposition 5.2. Let $n_c = n, \gamma > 0, \beta_0 \in [0, 1]$, suppose there exist $n \times n$
nonnegative-definite matrices $P, Q, \hat{P}$ satisfying

$$0 = A^T P + PA + R_1 - \tfrac{1}{2}[\delta^{-1}(1 + \beta_0)^2 - \gamma^{-1}(1 - \beta_0)]P\Sigma P, \tag{5..9}$$

$$0 = [A - Q\overline{\Sigma} + \tfrac{1}{4}\gamma^{-1}(1 - \beta_0)^2 \Sigma P]^T \hat{P} + \hat{P}[A - Q\overline{\Sigma} + \tfrac{1}{4}\gamma^{-1}(1 - \beta_0)^2 \Sigma P]$$
$$+ \tfrac{1}{8}\gamma^{-1}(1 - \beta_0)^3 \hat{P}\Sigma\hat{P} + \tfrac{1}{2}\delta^{-1}(1 + \beta_0)^2 P\Sigma P, \tag{5..10}$$

$$0 = [A + \tfrac{1}{4}(\gamma^{-1}(1 - \beta_0)^2 - \delta^{-1}(1 - \beta_0)(1 + \beta_0)^2)\Sigma(P + \hat{P})]Q$$
$$+ Q[A + \tfrac{1}{4}(\gamma^{-1}(1 - \beta_0)^2 - \delta^{-1}(1 - \beta_0)(1 + \beta_0)^2)\Sigma(P + \hat{P})]^T$$
$$+ V_1 - Q\overline{\Sigma}Q, \tag{5..11}$$

and let A_c, B_c, C_c, E_c be given by (4..16)-(4..19). Furthermore, suppose that
$(\tilde{A}, \tilde{C})$ is observable. Then $\tilde{P} = \begin{bmatrix} P + \hat{P} & -\hat{P} \\ -\hat{P} & \hat{P} \end{bmatrix}$ satisfies (2..8). Furthermore,
the equilibrium solution $\tilde{x}(t) \equiv 0$ of the closed-loop system (2..1), (2..2) is
asymptotically stable, and $\mathcal{D}$ defined by (2..9) is a subset of the domain of
attraction of the closed-loop system.

Proof. The proof is similar to the proof of Proposition 5.1 below with $n_c = n$
and $\Gamma = G^T = \tau = I$.

6. Numerical Algorithm

Here we adopt the numerical algorithm given by [19] to solve the matrix equations given by Propositions 3.2, 3.1, 4.2, 4.1, 5.2, and 5.1. The basic algorithm is demonstrated by means of Proposition 3.1.

Algorithm 6.1.
Step 1. Choose iteration number i_{max} and tolerance ε. Initialize $i = 0$ and $\tau = I_n$.
Step 2. Compute $P, Q, \hat{P}, \hat{Q}$ given by (3..9) - (3..12).

> Step 2.1. Solve the Riccati equation (3..9) to obtain P.
> Step 2.2. Solve equations (3..10) and (3..11) simultaneously to obtain $\hat{P}$ and Q. Starting with an initial choice of Q, solve (3..10) to obtain $\hat{P}$. Substitute P and $\hat{P}$ into (3..11) and solve for Q. Repeat the above process until $\hat{P}$ and Q converge.
> Step 2.3. Solve the Lyapunov equation (3..12) to obtain $\hat{Q}$.

Step 3. Use contragradient diagonalization to update τ. First, compute $S \in \mathbf{R}^{n \times n}$ such that

$$S^T \hat{P} S = S^{-1} \hat{Q} S^{-T} = D,$$

where $D \triangleq \mathrm{diag}(d_1, \ldots, d_n)$ and $d_1 > \cdots > d_n > 0$. Then construct τ as

$$\tau = S \hat{\tau} S^{-1},$$

where $\hat{\tau} \triangleq \mathrm{diag}(\hat{\tau}_1, \ldots, \hat{\tau}_n)$ is defined by

$$\hat{\tau}_j = 1, \qquad j \le n_c,$$
$$= d_j / d_{n_c}, \quad j > n_c.$$

Step 4. If $i = i_{max}$, go to Step 6. Otherwise, go to Step 5.
Step 5. If $(\mathrm{tr}\,(\hat{\tau}) - n_c)/n_c < \varepsilon$, go to Step 6. Otherwise, increment $i = i + 1$ and go to Step 2.
Step 6. Compute

$$\Gamma = [I_{n_c}\ 0]S^{-1}, \quad G = [I_{n_c}\ 0]S^T,$$

and calculate A_c, B_c, C_c using (3..14) - (3..16).

7. Numerical Examples

In this section, we reconsider the examples given in [15] to demonstrate the linear and nonlinear full-order and reduced-order compensators given by Propositions 3.2, 4.2, 5.2 and 3.1.

Example 7.1. To illustrate Proposition 3.2, consider the asymptotically stable system

$$\dot{x} = \begin{bmatrix} -0.03 & 1 & 0 \\ 0 & -0.03 & 1 \\ 0 & 0 & -0.03 \end{bmatrix} x + \begin{bmatrix} 0 \\ 0 \\ 1 \end{bmatrix} \sigma(u(t)),$$

$$y = \begin{bmatrix} 1 & 0 & 0 \end{bmatrix} x,$$

with the saturation nonlinearity $\sigma(u)$ given by

$$\begin{aligned} \sigma(u) &= u, & |u| < 4, \\ &= \text{sgn}(u)4, & |u| \geq 4. \end{aligned}$$

Choosing $R_1 = I_3, R_2 = 100, V_1 = I_3, V_2 = 1$, and $\beta_0 = 0.3, R_0 = \frac{50}{1-\beta_0}R_2 = 7142.9$, Algorithm 6.1 yields the linear controller (3..3), (3..4) with gains (3..25) - (3..27) given by

$$A_c = \begin{bmatrix} -2.8715 & 1.0000 & 0 \\ -3.6223 & -0.0300 & 1.0000 \\ -2.3579 & -0.0841 & -0.4114 \end{bmatrix}, \quad B_c = \begin{bmatrix} 2.8415 \\ 3.6223 \\ 2.3482 \end{bmatrix},$$

$$C_c = \begin{bmatrix} -0.0215 & -0.1866 & -0.8462 \end{bmatrix}.$$

By applying Remark 3.1, the set $\mathcal{D}$ is given by $\mathcal{D} = \{x_0 : x_0^T(P + \hat{P})x_0 < 6.0776 \times 10^4\}$, where

$$P + \hat{P} = 10^3 \times \begin{bmatrix} 1.1129 & -0.0101 & -1.2121 \\ -0.0101 & 1.7482 & -1.3314 \\ -1.2121 & -1.3314 & 6.6099 \end{bmatrix}.$$

To illustrate the closed-loop behavior let $x_0 = [-40 \ -25 \ 30]^T$ and $x_{c0} = [0 \ 0 \ 0]^T$, respectively. Note that $x_0^T(P + \hat{P})x_0 = 1.3708e + 07$, so that x_0 is not an element of $\mathcal{D}$. As can be seen in Figure 8.4, the closed-loop system consisting of the saturation nonlinearity and the LQG controller designed for the "unsaturated" plant exhibits limit cycle. However, the controller designed by Proposition 3.2 provides an asymptotically stable closed-loop system. The actual domain of attraction is thus larger than $\mathcal{D} \times \{0\}$. Figure 8.5 illustrates the control input $u(t)$ for the LQG controller with and without saturation as well as the output of the saturation nonlinearity $\sigma(u(t))$ for the LQG controller with saturation. Figure 8.6 shows the control $u(t)$ and saturated input $\sigma(u(t))$ for the controller obtained from Proposition 3.2. Comparing the results with the same example given in [15], it can be seen from Figure 8.7 that the chattering behavior of the saturated input $\sigma(u(t))$ for $0 \leq t \leq 0.5$ that occurred in [15] does not arise in the positive real approach. However, Figure 8.8 shows that the result of [15] yields less overshoot than Proposition 3.2. Although, for this open loop stable system, the given controller (A_c, B_c, C_c) does not provide global asymptotic stability, it does give faster response than globally stabilizing controllers such as $u(t) \equiv 0$.

Example 7.2. To illustrate Proposition 4.2, we reconsider Example 7.1 with $\beta_0 = 0.2$, and $R_1 = I_3, R_2 = 100, R_0 = \frac{200}{1-\beta_0} R_2, V_1 = I_3$, and $V_2 = 1$. Applying Algorithm 6.1 to Proposition 4.2 yields the nonlinear controller (4..1), (4..2) with gains (4..16) - (4..19) given by

$$A_c = \begin{bmatrix} -2.3553 & 1.0000 & 0 \\ -2.2733 & -0.0300 & 1.0000 \\ -0.9378 & -0.0903 & -0.5999 \end{bmatrix}, \quad B_c = \begin{bmatrix} 2.3253 \\ 2.2733 \\ 0.9301 \end{bmatrix},$$

$$C_c = \begin{bmatrix} -0.0105 & -0.1233 & -0.7785 \end{bmatrix}, \quad E_c = \begin{bmatrix} 0 \\ 0 \\ 0.6 \end{bmatrix}.$$

By applying Remark 3.1, the set $\mathcal{D}$ is given by $\mathcal{D} = \{x_0 : x_0^T(P + \hat{P})x_0 < 6.5384 \times 10^5\}$, where

$$P + \hat{P} = 10^4 \times \begin{bmatrix} 0.1034 & 0.0069 & -0.2504 \\ 0.0069 & 0.2318 & -0.2013 \\ -0.2504 & -0.2013 & 2.6019 \end{bmatrix}.$$

To illustrate the closed-loop behavior let $x_0 = [-40 \;\; -25 \;\; 30]^T$ and $x_{c0} = [0 \;\; 0 \;\; 0]^T$, respectively. Note that $x_0^T(P + \hat{P})x_0 = 3.5686 \times 10^7$, that is, x_0 is not in the set $\mathcal{D}$. As can be seen in Figure 8.9, the closed-loop system consisting of the saturation nonlinearity and the LQG controller designed for the "unsaturated" plant is unstable. However, the controller designed by Proposition 4.2 provides an asymptotically stable closed-loop system. The actual domain of attraction is thus larger than $\mathcal{D} \times \{0\}$. Figure 8.10 illustrates the control input $u(t)$ for the LQG controller with and without saturation as well as the output of the saturation nonlinearity $\sigma(u(t))$ for the LQG controller with saturation. Figures 8.11 shows the control $u(t)$ and saturated input $\sigma(u(t))$ for the controller obtained from Proposition 4.2.

Example 7.3. This example illustrates Proposition 5.2 for designing nonlinear controllers with integrators for tracking step commands. Consider the closed-loop system shown in Figure 8.12, where the plant $G(s) = 1/s^2$ and r is a step command. Let $G(s)$ and $G_c(s)$ have the realizations

$$\begin{bmatrix} \dot{x}_1 \\ \dot{x}_2 \end{bmatrix} = \begin{bmatrix} 0 & 1 \\ 0 & 0 \end{bmatrix} \begin{bmatrix} x_1 \\ x_2 \end{bmatrix} + \begin{bmatrix} 0 \\ 1 \end{bmatrix} \sigma(u),$$

$$y = \begin{bmatrix} 1 & 0 \end{bmatrix} \begin{bmatrix} x_1 \\ x_2 \end{bmatrix},$$

and

$$\dot{x}_c = A_c x_c + B_c q + E_c(\sigma(u) - u),$$
$$u = C_c x_c,$$

respectively. The saturation nonlinearity $\sigma(u)$ is given by

$$\sigma(u) = u, \qquad\qquad |u| < 0.3,$$
$$= \mathrm{sgn}(u)0.3, \quad |u| \geq 0.3.$$

To apply Theorem 2.1, we combine the plant $G(s)$ with an integrator state q to obtain the augmented plant

$$\begin{bmatrix} \dot{x}_2 \\ \dot{e} \\ \dot{q} \end{bmatrix} = \begin{bmatrix} 0 & 0 & 0 \\ -1 & 0 & 0 \\ 0 & 1 & 0 \end{bmatrix} \begin{bmatrix} x_2 \\ e \\ q \end{bmatrix} + \begin{bmatrix} 1 \\ 0 \\ 0 \end{bmatrix} \sigma(u(t)),$$

$$q = \begin{bmatrix} 0 & 0 & 1 \end{bmatrix} \begin{bmatrix} x_2 \\ e \\ q \end{bmatrix},$$

which has the form of (3..1), (3..2) with $x = \begin{bmatrix} x_2 \\ e \\ q \end{bmatrix}$. Choosing $\beta_0 = 0.2$,

$R_1 = I_3, R_2 = 100, V_1 = I_3, V_2 = 1$, and $R_0 = \frac{5000}{1-\beta_0}R_2$, Algorithm 6.1 applied to Proposition 5.2 yields the nonlinear controller (4..1), (4..2) with gains (4..16) - (4..19) given by

$$A_c = \begin{bmatrix} -0.4405 & 0.0441 & 0.6881 \\ -1.0000 & 0 & -2.0609 \\ 0 & 1.0000 & -2.2631 \end{bmatrix}, \quad B_c = \begin{bmatrix} -0.6859 \\ 2.0609 \\ 2.2631 \end{bmatrix},$$

$$C_c = \begin{bmatrix} -0.6007 & 0.0601 & 0.0030 \end{bmatrix}, \quad E_c = \begin{bmatrix} 0.6 \\ 0 \\ 0 \end{bmatrix}.$$

The set $\mathcal{D}$ is given by $\mathcal{D} = \{x_0 : x_0^T(P + \hat{P})x_0 < 1.0596 \times 10^6\}$, where

$$P + \hat{P} = 10^5 \times \begin{bmatrix} 4.5372 & 0.3640 & 0.2541 \\ 0.3640 & 0.2602 & 0.0000 \\ 0.2541 & 0.0000 & 0.0823 \end{bmatrix}.$$

To illustrate the closed-loop behavior let $r = 5, x_{20} = q_0 = 0, e_0 = r$, and $x_{c0} = [0 \ 0 \ 0]^T$, respectively. Note that $x_0^T(P + \hat{P})x_0 = 6.5055 \times 10^5$, so that x_0 is in the subset $\mathcal{D}$ of the guaranteed domain of attraction. As can be seen from Figure 8.13, the output y of the closed-loop system with LQG controller becomes oscillatory and has a large overshoot, while the output of the closed-loop system with the controller given by Proposition 5.2 shows satisfactory response. Figure 8.14 shows the control input $u(t)$ for the LQG controller with and without saturation as well as the output of the saturation nonlinearity $\sigma(u(t))$ for the LQG controller with saturation. For the same example considered in [15], the initial condition is *not* in the guaranteed domain of attraction. Hence for this example the estimate of the domain of attraction provided by the positive real result appears to be less conservative than that provided by the small gain result.

Example 7.4. To illustrate Proposition 3.1 for the reduced-order case, consider the asymptotically stable system given by

$$\dot{x} = \begin{bmatrix} -0.2 & 1 & 0 \\ 0 & -0.2 & 1 \\ 0 & 0 & -0.2 \end{bmatrix} x + \begin{bmatrix} 0 \\ 0 \\ 1 \end{bmatrix} \sigma(u(t)),$$

$$y = \begin{bmatrix} 1 & 0 & 0 \end{bmatrix} x,$$

with the saturation nonlinearity $\sigma(u)$ given by

$$\begin{aligned} \sigma(u) &= u, & |u| < 1, \\ &= \operatorname{sgn}(u), & |u| \geq 1. \end{aligned}$$

With the weighting matrices $R_1 = I_3, R_2 = 100, V_1 = I_3$, and $V_2 = 1$, we obtain a stable controller which has balanced controllability and observability gramians given by $\operatorname{diag}(82.62, 22.20, 2.07)$. Truncating the third state yields the second-order controller

$$A_c = \begin{bmatrix} -1.9522 & -1.5587 \\ 1.5587 & -0.0001 \end{bmatrix}, \quad B_c = \begin{bmatrix} 0.7804 \\ -0.0044 \end{bmatrix},$$

$$C_c = \begin{bmatrix} -0.7804 & -0.0044 \end{bmatrix},$$

To apply Proposition 3.1 we use the same weighting matrices and let $\beta_0 = 0.3, R_0 = \frac{50}{1-\beta_0} R_2 = 7142.9$. Solving (3..9) - (3..12) yields the linear controller (3..3), (3..4) with gains (3..14) - (3..16) given by

$$A_c = \begin{bmatrix} -1.4738 & -4.4503 \\ 0.4885 & -0.0095 \end{bmatrix}, \quad B_c = \begin{bmatrix} 1.4357 \\ 0.0031 \end{bmatrix},$$

$$C_c = \begin{bmatrix} -0.0896 & -0.0342 \end{bmatrix}.$$

By applying Remark 3.1, the set $\mathcal{D}$ is given by $\mathcal{D} = \{x_0 : x_0^T(P + \hat{P})x_0 < 6.5664 \times 10^3\}$, where

$$P + \hat{P} = \begin{bmatrix} 7.0138 & 5.7758 & -2.2916 \\ 5.7758 & 32.5589 & 82.1084 \\ -2.2916 & 82.1084 & 448.1960 \end{bmatrix}.$$

To illustrate the closed-loop behavior let $x_0 = \begin{bmatrix} -40 & -25 & 30 \end{bmatrix}^T$ and $x_{c0} = \begin{bmatrix} 0 & 0 \end{bmatrix}^T$, respectively. Note that $x_0^T(P + \hat{P})x_0 = 3.2884e + 05$, so that x_0

is not an element of $\mathcal{D}$. As can be seen in Figure 8.16, the controller designed by Proposition 3.1 converges to the origin faster than the closed-loop system consisting of the saturation nonlinearity and the reduced-order LQG controller designed for the "unsaturated" plant. The actual domain of attraction is thus larger than $\mathcal{D} \times \{0\}$. Figure 8.17 illustrates the saturation nonlinearity $\sigma(u(t))$ for the truncated LQG controller and the controller obtained from Proposition 3.1.

8. Conclusions

In this paper, we developed full- and reduced-order linear and nonlinear dynamic compensators based upon Theorem 2.1, which accounts for the saturation nonlinearity and provides a guaranteed domain of attraction by means of a positive-real type Riccati equation. Theorem 2.1 also provides a direct link between the guaranteed domain of attraction and the performance index. Controller gains were characterized by Riccati equations that were obtained by minimizing an LQG-type cost. The synthesis approach was demonstrated by numerical examples involving full- and reduced-order linear, and nonlinear dynamic compensators. A numerical algorithm based upon [19] was adopted for solving the coupled design equations. More sophisticated algorithms based upon homotopy methods can also be developed as in [20, 21].

References

1. D. K. Lindner, T. P. Celano, E. N. Ide, "Vibration Suppression Using a Proof-mass Actuator Operating in Stroke/Force Saturation," *J. Vibr. Acoustics,* Vol. 113, pp. 423-433, 1991.

2. P. C. Shrivastava and R. F. Stengel, "Stability Boundaries for Aircraft with Unstable Lateral-Directional Dynamics and Control Saturation," *J. Guidance Contr. Dyn.,* Vol. 12, pp. 62-70, 1989.

3. J. L. LeMay, "Recoverable and Reachable Zones for Control Systems with Linear Plants and Bounded Controller Outputs," *IEEE Trans. Autom. Contr.,* Vol. AC-9, pp. 346-354, 1964.

4. A. T. Fuller, "In-the-Large Stability of Relay and Saturating Control Systems with Linear Controllers," *Int. J. Contr.,* Vol. 10, pp. 457-480, 1969.

5. J. F. Frankena and R. Sivan, "A Non-linear Optimal Control Law for Linear Systems," *Int. J. Contr.,* Vol. 30, pp. 159-178, 1979.

6. E. P. Ryan, *Optimal Relay and Saturating Control System Synthesis,* Peregrinus, 1982.

7. I. Horowitz, "A Synthesis Theory for a Class of Saturating Systems," *Int. J. Contr.,* Vol. 38, pp. 169-187, 1983.

8. E. D. Sontag, "An Algebraic Approach to Bounded Controllability of Linear Systems," *Int. J. Contr.,* Vol. 39, pp. 181-188, 1984.

9. P. J. Campo and M. Morari, "Robust Control of Processes Subject to Saturation Nonlinearities," *Comp. Chem. Eng.,* Vol. 14, pp. 343-358, 1990.

10. Z. Lin and A. Saberi, "Semi-global Exponential Stabilization of Linear Systems Subject to 'Input Saturation' via Linear Feedbacks," *Sys. Contr. Lett.,* Vol. 21, pp. 225-239, 1993.

11. A. R. Teel, "Semi-global Stabilizability of Linear Null Controllable Systems with Input Nonlinearities," *Proc. Amer. Contr. Conf.,* pp. 947-951, Baltimore, MD, June, 1994.

12. A. R. Teel, "Semi-global Stabilizability of Linear Null Controllable Systems with Input Nonlinearities," *IEEE Trans. on Autom. Contr.,* Vol. 40, No. 1, pp. 96-100, 1995.

13. A. Saberi, Z. Lin, and A. R. Teel, "Control of Linear Systems with Saturating Actuators," *IEEE Trans. on Autom. Contr.,* Vol. 41, No. 3, pp. 368-378, 1996.

14. M. V. Kothare, P. J. Campo, M. Morari, and C. N. Nett, "A Unified Framework for the Study of Anti-Windup Designs," *Automatica,* Vol. 30, pp. 1869-1883, 1994.

15. F. Tyan and D. S. Bernstein, "Antiwindup Compensator Synthesis for Systems with Saturating Actuators," *Int. J. Robust Nonlinear Contr.,* Vol. 5, pp. 521-538, 1995.

16. R. L. Kosut, "Design of Linear Systems with Saturating Linear Control and Bounded States," *IEEE Trans. Autom. Contr.,* Vol. AC-28, pp. 121-124, 1983.

17. D. S. Bernstein and W. M. Haddad, "LQG Control With an H_∞ Performance Bound: A Riccati Equation Approach," *IEEE Trans. Autom. Contr.,* Vol. 34, pp. 293-305, 1989.

18. B. D. O. Anderson, "A System Theory Criterion for Positive Real Matrices," *SIAM J. Contr. Optim.,* Vol. 5, pp. 171-182, 1967.

19. S. W. Greeley and D. C. Hyland, "Reduced-Order Compensation: Linear-Quadratic Reduction Versus Optimal Projection," *J. Guidance Contr. Dyn.,* Vol. 11, pp. 328-335, 1988.

20. D. Zigic, L. T. Watson, E. G. Collins, Jr., and D. S. Bernstein, "Homotopy Methods for Solving the Optimal Projection Equations for the H_2 Reduced Order Model Problem," *Int. J. Contr.*, Vol. 56, pp. 173-191, 1992.
21. Y. Ge, L. T. Watson, E. G. Collins, and D. S. Bernstein, "Probability-One Homotopy Algorithms for Full and Reduced Order H^2/H^∞ Controller Synthesis," *Proc. IEEE Conf. Dec. Contr.*, pp. 2672-2677, Orlando, FL, December 1994.

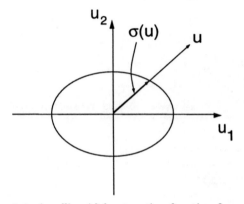

Fig. 8.1. An ellipsoidal saturation function for $m = 2$

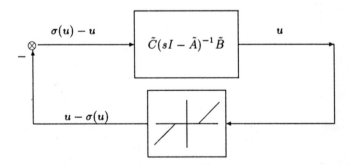

Fig. 8.2. Closed-loop system with a deadzone nonlinearity.

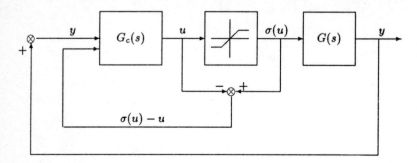

Fig. 8.3. Closed-loop system with a nonlinear observer-based antiwindup controller

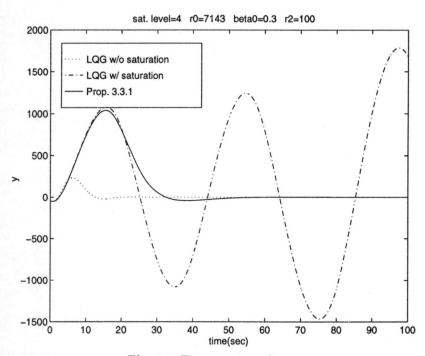

Fig. 8.4. Time response of y

153

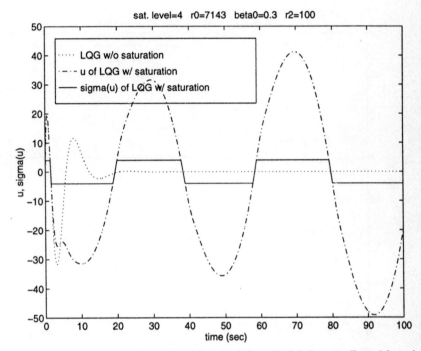

Fig. 8.5. Control effort u and saturated input $\sigma(u)$ of the LQG controller with and without saturation

154

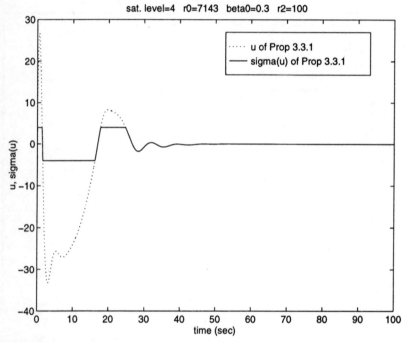

Fig. 8.6. Control effort u and saturated input $\sigma(u)$ using Proposition 3.1

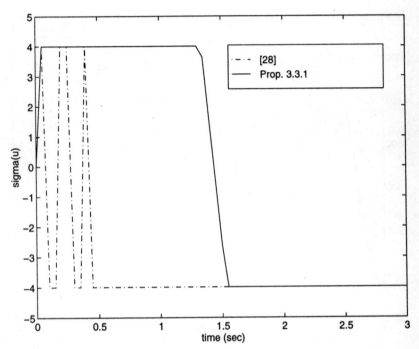

Fig. 8.7. Comparison of the saturated input $\sigma(u(t))$ of Example 7.1

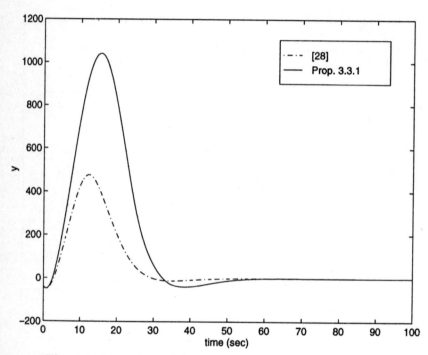

Fig. 8.8. Comparison of the time response of y for Example 7.1

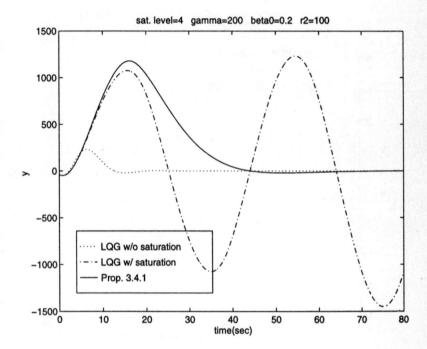

Fig. 8.9. Time response of y

158

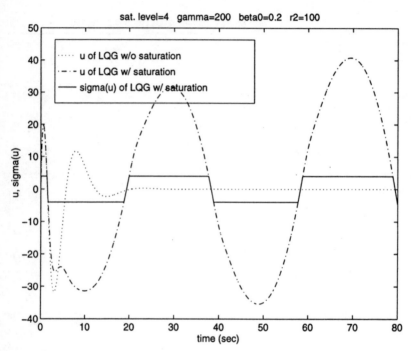

Fig. 8.10. Control effort u and saturated input $\sigma(u)$ of the LQG controller with and without saturation

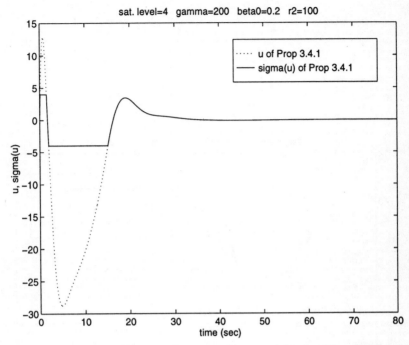

Fig. 8.11. Control effort u and saturated input $\sigma(u)$ using Proposition 4.1

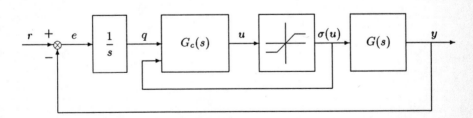

Fig. 8.12. Block diagram for Example 7.3

160

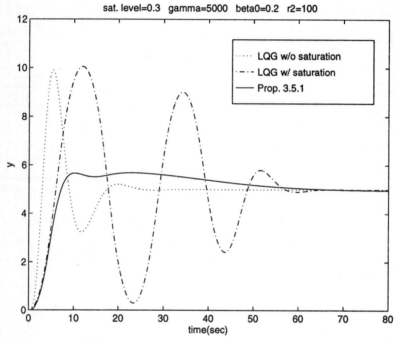

Fig. 8.13. Time response of y with $r = 5$

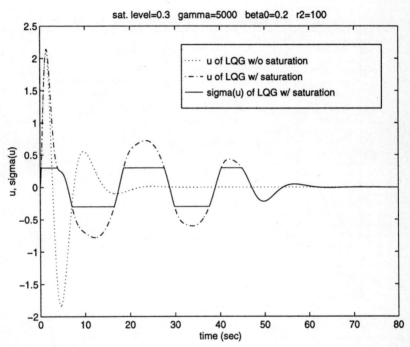

Fig. 8.14. Control effort u and saturated input $\sigma(u)$ of the LQG controller with and without saturation

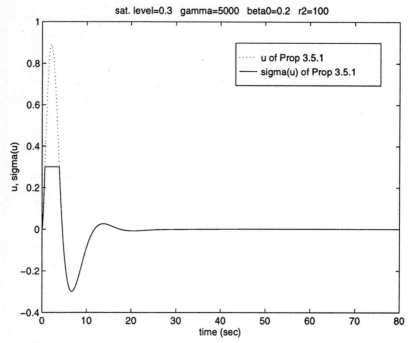

Fig. 8.15. Control effort u and saturated input $\sigma(u)$ using Proposition 5.1

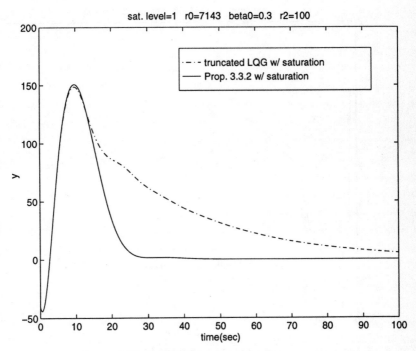

Fig. 8.16. Time response of y

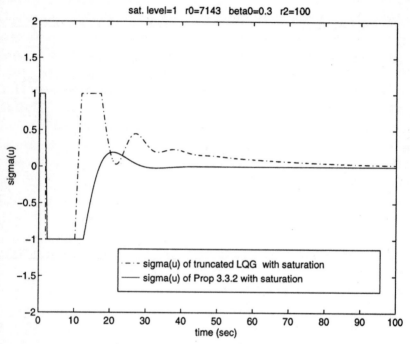

Fig. 8.17. Saturated input $\sigma(u)$ of the truncated LQG controller and Proposition 3.2

Chapter 9. Quantifier Elimination Approach to Frequency Domain Design

Peter Dorato, Wei Yang, and Chaouki T. Abdallah

Department of Electrical and Computer Engineering
University of New Mexico
Albuquerque, NM 87131-1356, USA
peter@eece.unm.edu

1. Introduction

Many robust feedback design problems, with frequency domain specifications, can be reduced to the satisfaction of inequalities of the form,

$$F_i(\omega, p, q) > 0, i = 1, 2, ..., m > 0 \qquad (1.1)$$

where ω represents the frequency variable, the vector p represents uncertain plant parameters, the vector q represents controller design parameters, and where the function F_i is a *multivariable polynomial* function of its arguments. A multivariable polynomial function is a function which is a polynomial in any given variable when all the other variables are held fixed. For quantifier elimination algorithms, the further assumption that the coefficients in F_i must be integers is required. Since real numbers can always be closely approximated by rational numbers, this is not a serious constraint.

Given a plant transfer function $G(s, p)$, and a controller with transfer function $C(s, q)$ in a unit-feedback structure, the requirement that the transfer function between reference input and *control input* be constrained to have a magnitude less than a given value, α_U, may be written

$$\left| \frac{C(j\omega, q)}{1 + C(j\omega, q)G(j\omega, p)} \right| < \alpha_U, \ 0 \le \omega \le \omega_1 \qquad (1.2)$$

By squaring the magnitude and clearing fractions, equation (1.2) takes on the form $F_1(\omega, p, q) > 0$, where the the function F_1 is polynomial in its arguments as long as the components of the vectors p and q enter polynomially in the coefficients of the polynomials in the transfer functions G and C (It is assumed that all transfer function are rational functions of the Laplace variable s). Similarly tracking error specifications can be met, in the frequency domain, by requiring the satisfaction of the inequality

$$\left| \frac{1}{1 + C(j\omega, q)G(j\omega, p)} \right| < \alpha_T, \ 0 \le \omega \le \omega_2 \qquad (1.3)$$

This inequality can also be reduced to an inequality of the form $F_2(\omega, p, q) > 0$. Finally stability of the closed-loop system is guaranteed, via the Routh Hurwitz test, by the satisfaction of further inequalities of the form $F_i(p, q) > 0$.

This polynomial-inequality formulation is especially attractive for control problems where there are no general analytical design algorithms and where, for practical reasons, one would like to have the simplest possible compensator. An example of this is the static *output-feedback stabilization problem*, which is the problem of finding a "zero-order" compensator, or what is commonly referred to *simple proportional feedback*, such that the closed-loop system is stabilized. Proportional feedback is the simplest possible type of feedback that can be used, yet this problem remains an open analytical problem [13].

In this paper, we propose the use of Quantifier Elimination (QE) methods to solve some of these control problems. In section 2 we present the reduction of many control problems to QE problems. Section 3 reviews some of the algorithms and software used to solve the QE problems, while section 4 contains examples of applying such software to some control problems. Our conclusions are presented in section 5.

2. Reduction to a Quantifier Elimination (QE) Problem

From the discussion in section 1 it follows that in the frequency domain, the problem of *control of uncertain systems with bounded control inputs* can be reduced to the satisfaction of inequality constraints of the form given in (1.1) with logic quantifiers of the form "for all ω" and "for all p" over given ranges of ω and p. Typically the variables in the polynomials are real and are related to plant (controlled system) and compensator (controller) parameters. The final design objective is to obtain quantifier-free formulas for the compensator parameters or, for the existence problem, to obtain a "true" or "false" output. As mentioned in section 1, QE methods are especially attractive for control problems where there are no general analytical design algorithms and where, for practical reasons, one would like to have the simplest possible compensator. With the notation $\forall$ for the "for all" logic quantifier, the design problem is the reduced to the elimination of the quantifiers $\forall$ in the logic formula

$$\forall(p)\forall(\omega)[F_1 > 0 \wedge F_2 > 0 \wedge ...] \tag{2.1}$$

where $\wedge$ denotes the logic "and" operator. This elimination produces a quantifier-free Boolean formula, $\Psi(q)$, (inequalities with logic "or" and "and" conjunctives) in the components of the design vector q. This formula may then be used to specify an admissible set of design parameters. The quantifier "there exists", denoted $\exists$, may be added to the formula in (2.1) to settle to question of existence of a controller that can robustly meet all the specifications.

3. QE Algorithms and Software

In this section, we review the general QE problem and introduce the software package QEPCAD which we use to solve our control problems. A more detailed treatment may be found in [3, 14].

Given the set of polynomials with integer coefficients $P_i(X, Y)$, $1 \leq i \leq s$ where X represents a k dimensional vector of quantified real variables and Y represents a l dimensional vector of unquantified real variables, let $X^{[i]}$ be a block of k_i quantified variables, Q_i be one of the quantifiers $\exists$ (there exists) or $\forall$ (for all), and let $\Phi(Y)$ be the quantified formula

$$\Phi(Y) = (Q_1 X^{[1]}, ..., Q_w X^{[w]}) F(P_1, ..., P_s), \qquad (3.1)$$

where $F(P_1, ..., P_s)$ is a quantifier free *Boolean* formula, that is a formula containing the Boolean operators $\wedge$ (and) and $\vee$ (or), operating on *atomic predicates* of the form $P_i(Y, X^{[1]}, ..., X^{[w]}) \geq 0$ or $P_i(Y, X^{[1]}, ..., X^{[w]}) > 0$ or $P_i(Y, X^{[1]}, ..., X^{[w]}) = 0$. We can now state the general quantifier elimination problem

General Quantifier Elimination Problem: Find a quantifier-free Boolean formula $\Psi(Y)$ such that $\Phi(Y)$ is true if and only if $\Psi(Y)$ is true.

In control problems, the unquantified variables are generally the compensator parameters, represented by the parameter vector $Y = q$, and the quantified variables are the plant parameters, represented by the plant parameter vector p, and the frequency variable ω. Uncertainty in plant parameters are characterized by quantified formulas of the type $\forall(p_i)$ $[\underline{p_i} \leq p_i \leq \overline{p_i}]$ where $\underline{p_i}$ and $\overline{p_i}$ are rational numbers. *The quantifier-free formula $\Psi(q)$ then represents a characterization of the compensator design.*

An important special problem is the QE problem with no unquantified variables (free variables), i.e. $l = 0$. This problem is referred to as the *General Decision Problem*.

General Decision Problem: With no unquantified variables, i.e. $l = 0$, determine if the quantified formula given in 3.1 is true or false.

The general decision problem may be applied to the problem of *existence* of compensators that meet given specifications, in which case an "existence" quantifier is applied to the compensator parameter q. Algorithms for solving general QE problems were first given by Tarski [14] and Seidenberg [11], and are commonly called Seidenberg-Tarski decision procedures. Tarski showed that QE is solvable in a finite number of "algebraic" steps, but his algorithm and later modifications are exponential in the size of the problem. Researchers in control theory have been aware of Tarski's results and their applicability to control problems since the 1970's [2], but the complexity of the computations and lack of software limited their applicability. Later, Collins [4] introduced a theoretically more efficient QE algorithm that uses a cylindrical algebraic decomposition (CAD) approach. However, this algorithm was not capable of effectively handling nontrivial problems. More recently Hong [8], Collins and

Hong [5], Hong [9] have introduced a significantly more efficient partial CAD QE algorithm.

The Cylindrical Algebraic Decomposition (CAD) algorithm, has been developed (See reference [5]) for the computer elimination of quantifiers on polynomial-function inequalities. This algorithm requires a *finite number* of "algebraic" operations. However the number of operations is still doubly exponential in the number of variables, so that only problems of modest complexity can actually be computed. See reference [3] for a discussion of computational complexity in quantifier elimination. A software package called QEPCAD (Quantifier Elimination by Partial Cylindrical Algebraic Decomposition) has been developed for the solution of quantifier elimination problems (H. Hong, Institute for Symbolic Computation, Linz, Austria). An excellent introduction to quantifier elimination theory and its applications to control system design may be found in the monograph of Jirstrand [12].

In the examples that follow we use the software package QEPCAD to solve some simple robust control-effort problems. It should be noted that numerical techniques can also be used to "eliminate quantifiers". For example in [7], Bernstein polynomial methods are used for this purpose. Indeed numerical techniques may be applicable to more complex problem than those that can be handled by QE algorithm. However numerical techniques generally require a priori bounds on design parameter range, and are also limited by problem size. A major advantage of QE algorithm is that they require no approximations or a priori parameter ranges. Nevertheless it is important to do as much "hand" reduction as possible before using QE software in order to overcome the computational complexities inherent in QE algorithms.

4. Examples

Example 4.1. The example here is taken from reference [7]. Consider a plant with transfer function,

$$G(s,p) = \frac{p_1}{1 - s/p_2}, \quad 0.8 \le p_{1,2} \le 1.25$$

with a simple proportional feedback controller,

$$C(s,q) = q_1$$

The specification data is given by:

$$\alpha_U = 20, \ \alpha_T = 0.2, \ \omega_2 = 2$$

With some computation, the control input specification results in the function,

$$F_1(\omega, p, q) = (400 - q_1^2)\omega^2 + (p_2)^2(400(1 + p_1 q_1)^2 - q_1^2)$$

the tracking specification results in the function,

$$F_2(\omega, p, q) = -24w^2 + (p_2)^2((1 + p_1q_1)2 - 25)$$

and the robust stability specification results in the function,

$$F_3(p, q) = -p_2(1 + p_1p_2).$$

The robust bounded input design problem is then reduced to the problem of eliminating the $\forall$ quantifiers in the quantified expression

$$\forall(16 \le 20p_{1,2} \le 25)\forall(\omega)\forall(0 \le \omega_1 \le 2)[F(\omega, p, q)]$$

where

$$F(\omega, p, q) = [F_1(\omega, p, q) > 0 \wedge F_2(\omega_1, p, q) > 0 \wedge F_3(p, q) > 0].$$

See figure 4.1 for QEPCAD input file for this example. QEPCAD software produces the quantifier-free formula,

$$\Psi(q_1) = (q_1 + 20 \ge 0) \wedge (8q_1^2 + 20q_1 - 2175 > 0) \wedge (5q_1 - 16 \le 0)$$

from which one can deduce, by computing roots of single-variable polynomials, the following range of acceptable design parameter values

$$-20 \le q_1 < -17.7895.$$

```
(q1,p1,p2,w1,w2)
1
(A p1) (A p2) (A w1) (A w2)
[
          [16 <= 20 p1 /\ 20 p1 <= 25 /\
           16 <= 20 p2 /\ 20 p2 <= 25
           /\ 0 <= w1 /\ w1 <= 2]
     ==>
          [p2 (1+p1 q1) < 0 /\
           -24 w1^2 + p2^2 ((1+ q1 p1)^2 - 25) > 0 /\
           (400-q1^2) w2^2 + p2^2 (400 (1+p1 q1)^2 - q1^2) > 0]
].
go
go
go
go
```

Fig. 4.1. QEPCAD Input File for Example 4.1

Example 4.2. This problem is taken from reference [6]. Consider a plant with transfer function,

$$G(s, p_1) = \frac{1}{s + p_1}, \ p_1 = \pm 1$$

with a simple PI feedback controller,

$$C(s, q_1, q_2) = q_1 + q_2/s.$$

The specification data is given by

$$\alpha_U = 5, \ \omega_1 = \infty.$$

With some computations, the specifications result in the functions

$$
\begin{aligned}
F_1(\omega, q_1, q_2) &= q_1 - 1 \\
F_2(\omega, q_1, q_2) &= q_2 \\
F_3(\omega, q_1, q_2) &= (5 - q_1^2)\omega^4 + (5((1 + q_1)^2 - 2q_2) - (q_2^2 + q_1^2))\omega^2 + 4q_2^2 \\
F_4(\omega, q_1, q_2) &= (5 - q_1^2)\omega^4 + (5((-1 + q_1)^2 - 2q_2) - (q_2^2 + q_1^2))\omega^2 + 4q_2^2
\end{aligned}
$$

which must all be positive. The first thing we explore is the existence of a compensator. The existence question may be explored by putting existence quantifier on both q_1 and q_2,

$$\forall(\omega)\exists(q_1, q_2)[F(\omega, q_1, q_2)]$$

where

$$
\begin{aligned}
F(\omega, q_1, q_2) = \ &[F_1(\omega, q_1, q_2) > 0 \wedge F_2(\omega, q_1, q_2) > 0 \wedge \\
&F_3(\omega, q_1, q_2) > 0 \wedge F_4(\omega, q_1, q_2) > 0].
\end{aligned}
$$

QEPCAD software returns a "yes" to the existence question. To find a quantifier free formula for the compensator parameters, the "$\forall$" quantifier should be eliminated from the expression $\forall(\omega)[F(\omega, q_1, q_2)]$. Dealing with four inequalities requires too much computer time with QEPCAD software. To reduce computer time the problem the problem is reformulated as two expressions with three inequalities, i.e.

$$\forall(\omega)[F_1(\omega, q_1, q_2) > 0 \wedge F_2(\omega, q_1, q_2) > 0 \wedge F_3(\omega, q_1, q_2) > 0]$$

where QEPCAD produces the quantifier formula $\Psi_1(q_1, q_2)$,

```
[ [ q1 - 1 > 0 /\ q1^2 - 5 <= 0 /\ q2 > 0 /\
q2^2 + 10 q2 - 4 q1^2 - 10 q1 - 5 <= 0 ] \/
[ q1 - 1 > 0 /\ q2 >= 0 /\
q2^4 + 20 q2^3 + 8 q1^2 q2^2 - 20 q1 q2^2 + 10 q2^2
- 80 q1^2 q2 - 200 q1 q2 - 100 q2 + 16 q1^4 + 80 q1^3
+ 140 q1^2 + 100 q1 + 25 <= 0 ] ]
```

and
$$\forall(\omega)[F_1(\omega, q_1, q_2) > 0 \wedge F_2(\omega, q_1, q_2) > 0 \wedge F_4(\omega, q_1, q_2) > 0]$$

where QEPCAD produces the quantifier formula $\Psi_2(q_1, q_2)$,

```
[ [ q1 - 1 >= 0 /\ q1^2 - 5 <= 0 /\ q2 > 0 /\
q2^2 + 10 q2 - 4 q1^2 + 10 q1 - 5 <= 0 ] \/
[ q1 - 1 >= 0 /\ q2 > 0 /\
q2^4 + 20 q2^3 + 8 q1^2 q2^2 + 20 q1 q2^2 + 10 q2^2
- 80 q1^2 q2 + 200 q1 q2 - 100 q2 + 16 q1^4 - 80 q1^3
+ 140 q1^2 - 100 q1 + 25 <= 0 ]].
```

The quantifier free formula for original problem is then $\Psi(q_1, q_2) = (\Psi_1(q_1, q_2) \wedge \Psi_2(q_1, q_2))$. The resulting "Boolean" formula in two variables q_1 and q_2 is rather complicated. We would like the results displayed in the form of ranges in design parameters. For this we obtain first an interval q_1 by formulating the following QE question

$$(\exists q_2)[\Psi(q_1, q_2)]$$

and eliminates the quantifier on q_2. QE produces the results below,

```
[ [ [ 4 q1^2 - 10 q1 + 5 > 0 /\
16 q1^4 - 80 q1^3 + 165 q1^2 - 100 q1 - 100 <= 0 ] \/
[ q1^2 - 5 <= 0 /\
16 q1^8 - 80 q1^7 + 180 q1^6 + 200 q1^5 + 1400 q1^4 +
1500 q1^3 - 10500 q1^2 - 2500 q1 + 625 >= 0 ] ] /\
[ [ q1 - 1 > 0 /\ q1^2 - 5 <= 0 ] \/ q1 - 1 > 0 ] /\
[ [ q1 - 1 >= 0 /\ q1^2 - 5 <= 0 ] \/ q1 - 1 >= 0 ] ]
```

from which one can reduce, by computing roots of single-variable polynomials, the following range: $1 < q_1 < 2.2361$. Then we discretize q_1 within this range and plug into the quantifier-formula $\Psi(q_1, q_2)$ a particular discretized value of q_1 to find what the acceptable range of q_2 is for this particular discretized value of q_1. The solution for this problem is represented by the table 4.1, for five discrete values of q_1. The notation $(a, b]$ in the table produces that q_2 is in the range, $a < q_2 \leq b$.

Table 4.1. Design parameter regions for example 4.2

Design parameter q_1	Design parameter q_2
1.2	(0, 3.6996]
1.5	(0, 3.9588]
1.7	(0, 4.0654]
2.0	(0, 0.1623]
2.2	(0, 0.2721]

5. Conclusion

The design, in the frequency domain, of robust feedback systems with bounded control effort can be reduced to quantifier elimination problem. However due to computational complexities, only problems of modest size can be solved. Nevertheless it may be possible to solve some practical problems where no analytic design procedures exist.

References

1. C. Abdallah, P. Dorato, W. Yang, R. Liska, and S. Steinberg, *Applications of quantifier elimination theory to control system design*, 4th IEEE Mediterranean Symposium on Control & Automation, Chania, Crete, Greece, June 10-14, 1996.
2. B.D.O. Anderson, N.K. Bose, and E.I. Jury, *Output feedback and related problems-Solution via Decision methods*, IEEE Trans. on Automatic Control, AC-20, pp.53-65, 1975.
3. S. Basu, R. Pollack, and M.F. Roy, *On the combinatorial and algebraic complexity of quantifier elimination*, Proc. 35th Symposium on Foundations of Computer Science, Santa Fe, NM, pp. 632-641, 1994.
4. G. E. Collins, *Quantifier Elimination in the Elementary Theory of Real Closed Fields by Cylindrical Algebraic Decomposition*, Lecture Notes in Computer Science, Spring Verlag, Berlin, **Vol. 33**, pp. 134-183, 1975.
5. G.E. Collins and H. Hong, *Partial cylindrical algebraic decomposition for quantifier elimination*, J. Symbolic Computation, 12, pp. 299-328, 1991.
6. P. Dorato, W. Yang, and C. Abdallah, *Robust multi-objective feedback design by quantifier elimination*, submitted to J. Symbolic Computation, 1996.
7. G.Fiorio, S. Malan, M.Milanese, and M. Taragna, *Robust Performance Design of Fixed Structure Controller with Uncertain Parameters*, Proc. 32nd IEEE Conf. on Decision and Control, San Antonio, TX, pp. 3029-3031.
8. H. Hong, *Improvements in CAD-based Quantifier Elimination*, Ph.D Thesis, The Ohio State University, 1990.
9. H. Hong, *Simple Solution Formula Construction in Cylindrical Algebraic Decomposition based Quantifier Elimination*, ISSAC'92, International Symposium on Symbolic and Algebraic Computation, July 27-29, Berkeley, California (Editor P.S. Wang), ACM Press, New York, pp. 177-188, 1992.
10. R. Liska and S. Steinberg, *Applying Quantifier Elimination to Stability Analysis of Difference Schemes*, The Computer Journal, **vol. 36, No. 5**, pp. 497-503, 1993.
11. A. Seidenberg, *A New Decision Method for Elementary Algebra*, Annals of Math., **60**, pp. 365-374, 1954.
12. M. Jirstrand, *Algebraic methods for modeling and design in control*, Linköping Studies in Science and Technology, Thesis no. 540, Linköping University, 1996.
13. V.L. Syrmos, C.T. Abdallah, P.Dorato, and K. Grigoriadis, *Static Output Feedback: A Survey*, Scheduled for publication in Automatica, Feb., 1997.
14. A. Tarski, *A Decision Method for Elementary Algebra and Geometry*, 2nd Ed., Berkeley, University of California Press, 1951.

Chapter 10. Stabilizing Feedback Design for Linear Systems with Rate Limited Actuators*

Zongli Lin[1], Meir Pachter[2], Siva Banda[3], Yacov Shamash[4]

[1] Dept. of Applied Math. & Stat. SUNY at Stony Brook Stony Brook, NY 11794-3600
[2] Dept. of Elect.& Comp. Sci. Air Force Inst. of Tech. Wright-Patterson AFB, OH 45433
[3] Flight Dynamics Dir. (WL/FGIC) Wright Laboratory Stony Brook, NY 11794-3600
[4] College of Engr. & Applied Sci. SUNY at Stony Brook Stony Brook, NY 11794-2200

1. Introduction

Every physical actuator is subject to constraints. These constraints include both position and rate saturation. In the past few years there has been much interest concerning stabilization of linear systems with position saturating actuators, resulting in several promising design techniques. In this paper, we will recourse to the low-and-high gain (LHG) design technique ([7]) and the piecewise linear LQ control (PLC) design technique ([8]). Additional related work might be found in these two papers and in [2].

While actuator position saturation has been addressed in the recent literature, few design techniques are currently available to deal with actuator rate saturation. However, actuator rate saturation often presents a more serious challenge to control engineers, especially flight control engineers. It is known ([1]) that actuator rate saturation could induce a considerable phase-lag. Such phase-lag associated with rate saturation has a destabilizing effect. For example, investigators have identified rate saturation as a contributing factor to the recent mishaps of YF-22 [3] and Gripen [4] prototypes and the first production Gripen [6]. For further discussion on the destabilizing effect of actuator rate saturation, see [1].

The objective of this paper is to propose a method of designing stabilizing feedback control laws for linear systems taking into account the effect of actuator rate saturation. The proposed design method views the problem of stabilization with rate saturating actuators as a problem of robust stabilization with position saturating actuators in the presence of input additive uncertainties. The state feedback law is then designed for stabilization with position saturating actuators. Our state feedback design combines the two recently developed design techniques, the PLC and the LHG design techniques.

* This work was conducted while the first author was participating in the 1996 AFOSR summer faculty research program. He acknowledges the support of AFOSR.

It inherits the advantages of the both design techniques, while avoiding their disadvantages. Thus, the exact knowledge of the dynamics of the actuators will not be needed and the actuator disturbances can be rejected. In particular, in the LHG design, a low gain feedback law is first designed in such a way that the actuator does not saturate in position and the the closed-loop system remains linear. The gain is chosen low to enlarge the region in which the closed-loop system remains linear and hence enlarge the domain of attraction of the closed-loop system. Then, utilizing an appropriate Lyapunov function for the closed-loop system under this low gain feedback control law, a linear high gain feedback control law is constructed and added to the low gain feedback control to form the final LHG feedback control law. Such a linear low-and-high gain feedback control law speeds up the transient response for the state in a certain subspace of the state space and is capable of stabilizing the system in the presence of input-additive plant uncertainties and rejecting arbitrarily large bounded input-additive disturbances. The disadvantage of this control law is that the transient response for the state outside that subspace of the state space remains that of the low gain feedback, which is typically sluggish (due to low feedback gain for a large domain of attraction). On the other hand, the aim of the PLC scheme is to increase the feedback gain piecewisely while adhering to the input bound as the trajectories converge toward the origin. Such a design results in fast transient speed for all states. However, it lacks robustness to large uncertainties and the ability of rejecting arbitrarily large bounded disturbances.

The remainder of the paper is organized as follows. In Section 2., we precisely formulate our problem. Section 3. provides the design algorithm and proves that the proposed algorithm results in feedback laws that solve the problem formulated in Section 2.. In Section 4., an F-16 class fighter aircraft model is used to demonstrate the effectiveness of the proposed design algorithm. Concluding remarks are in Section 5..

2. Problem Formulation

Consider the linear dynamical system

$$\begin{cases} \dot{x} = Ax + Bv, & x(0) \in \mathcal{X} \subset \mathbb{R}^n \\ \dot{v} = \text{sat}_\Delta(-T_1 v + T_2 u + d), & v(0) \in \mathcal{V} \subset \mathbb{R}^m \end{cases} \qquad (2.1)$$

where $x \in \mathbb{R}^n$ is the plant state, $v \in \mathbb{R}^m$ is the actuator state and input to the plant, $u \in \mathbb{R}^m$ is the control input to the actuators, for $\Delta = (\Delta_1, \Delta_2, \cdots, \Delta_m)$, $\Delta_i > 0$, the function $\text{sat}_\Delta : \mathbb{R}^m \to \mathbb{R}^m$ is the standard saturation function that represents actuator rate saturation, i.e., $\text{sat}_\Delta(v) = [\text{sat}_{\Delta_1}(v_1), \text{sat}_{\Delta_2}(v_2), \cdots, \text{sat}_{\Delta_m}(v_m)]$, $\text{sat}_{\Delta_i}(v_i) = \text{sign}(v_i) \min\{\Delta_i, |v_i|\}$, the positive definite diagonal matrices $T_1 = \text{diag}(\tau_{11}, \tau_{12}, \cdots, \tau_{1m})$ and $T_2 = \text{diag}(\tau_{21}, \tau_{22}, \cdots, \tau_{2m})$ represent the "time constants" of the actuators and are not precisely known, and finally $d : \mathbb{R}_+ \to \mathbb{R}^m$ are the disturbance signals appearing at the input of the actuators.

We also make the following necessary assumptions on the system.

Assumption 2.1. The pair (A, B) is stabilizable;

Assumption 2.2. The nominal value of T_1 and T_2 is known and is given by $T^* = \text{diag}(\tau_1^*, \tau_2^*, \cdots, \tau_m^*)$ with $\tau_i > 0$. There exist known matrices $\mu_1 = \text{diag}(\mu_{11}, \mu_{12}, \cdots, \mu_{1m})$, $\mu_2 = \text{diag}(\mu_{21}, \mu_{22}, \cdots, \mu_{2m})$, $|\mu_{ij}| \leq 1$, and $\nu_1 = \text{diag}(\nu_{11}, \nu_{12}, \cdots, \nu_{1m})$, $\nu_2 = \text{diag}(\nu_{21}, \nu_{22}, \cdots, \nu_{2m})$, $|\nu_{ij}| \geq 1$, such that, $\mu_1 T^* \leq T_1 \leq \nu_1 T^*$ and $\mu_2 T^* \leq T_2 \leq \nu_2 T^*$.

Assumption 2.3. The disturbance is uniformly bounded by a known (arbitrarily large) constant D, i.e., $\|d(t)\| \leq D, \forall t \geq 0$.

Before stating our problem, we make the following preliminary assumption.

Definition 2.1. *The data* $(D, \mathcal{W}_0)$ *is said to be* admissible for state feedback *if D is a nonnegative real number, and $\mathcal{W}_0$ is a subset of* $\mathbb{R}^{n+m}$ *which contains the origin as an interior point.*

The objective of this paper is to provide a design algorithm for solving the following problem.

Problem 2.1. Given the data $(D, \mathcal{W}_0)$, admissible for state feedback, find, if possible, a feedback law $u = F(x, v)$, such that the closed-loop system satisfies,

1. if $D = 0$, the point $x = 0$ is locally asymptotically stable and $\mathcal{X} \times \mathcal{V}$ is contained in its basin of attraction;
2. if $D > 0$, every trajectory starting from $\mathcal{X} \times \mathcal{V}$ enters and remains in $\mathcal{W}_0$ after some finite time.

3. A Combined PLC/LHG Design Algorithm

As stated earlier, the proposed design algorithm is a combination of the piecewise linear LQ control [8] and the low-and-high gain feedback [7]. Naturally we organize this section as follows. Subsections 3.1 and 3.2 respectively recapitulate the PLC and the LHG design techniques. Subsection 3.3 presents the proposed combined PLC/LHG design algorithm. Finally, in Subsection 3.4, the proposed design algorithm is shown to solve Problem 2.1.

3.1 Piecewise Linear LQ Control Design

Consider the linear dynamical system subject to actuator position saturation,

$$\dot{x} = Ax + B\text{sat}_\Delta(u), \quad x(0) \in \mathcal{X} \subset \mathbb{R}^n, \, u \in \mathbb{R}^m \tag{3.1}$$

where the saturation function $\text{sat}_\Delta : \mathbb{R}^m \to \mathbb{R}^m$ is as defined in Section 2.,
and the pair (A, B) is assumed to be stabilizable.

The PLC design is based on the following LQ algebraic Riccati equation,

$$A'P + PA - PBR^{-1}B'P + I = 0 \tag{3.2}$$

where $R = \text{diag}(\epsilon) = \text{diag}(\epsilon_1, \epsilon_2, \cdots, \epsilon_m)$, $\epsilon_i > 0$, are the design parameters
to be chosen later.

Key to the PLC scheme is the notion of invariant sets. A nonempty subset
of ε in $\mathbb{R}^n$ is positively invariant if for a dynamical system and for any
initial condition $x(0) \in \varepsilon$, $x(t) \in \varepsilon$ for all $t \geq 0$. For the closed-loop system
comprising of the system (3.1) and the LQ control $u = -R^{-1}B'Px$, simple
Lyapunov analysis shows that the Lyapunov level set

$$\varepsilon(P, \rho) = \{x : x'Px \leq \rho\}, \, \forall \rho > 0$$

is an invariant set, provided that saturation does not occur for all $x \in \varepsilon(P, \rho)$.
To avoid the saturation from occurring, while fully utilizing the available
control capacity, for a given ρ, $\epsilon = (\epsilon_1, \epsilon_2, \cdots, \epsilon_m)$ will be chosen to be the
largest such that

$$|u_i| = \left| \frac{1}{\epsilon_i} B_i' Px \right| \leq \Delta_i, \, \forall x \in \varepsilon(P, \rho)$$

where B_i is the ith column of matrix B and u_i is the ith element of u. The
existence and uniqueness of such an ϵ are established in [8] and an algorithm
for computing such an ϵ is also given in [8]. More specifically, it is shown
through the existence of a unique fixed point that the following iteration
converges from any initial value to the desired value of ϵ,

$$\epsilon_{n+1} = \sqrt{\rho}\Phi(\epsilon_n) \tag{3.3}$$

where

$$\Phi(\epsilon) = [\phi_1(\epsilon), \phi_2(\epsilon), \cdots, \phi_m(\epsilon)]'$$

and for each $i = 1$ to m,

$$\phi_i(\epsilon) = \frac{1}{\Delta_i}\sqrt{B_i'P(\epsilon)B_i}$$

The aim of the PLC scheme is to increase the feedback gain piecewisely
while adhering to actuator bounds as the trajectories converge towards the
origin. This is achieved by constructing nested level sets, $\varepsilon_0, \varepsilon_1, \cdots, \varepsilon_N$, in
such a way that as the trajectories traverse successively the surface of each

ε_i and the control law is switched to higher and higher gains as each surface is crossed.

The procedure in designing a PLC law is as follows. Given the set of initial conditions $\mathcal{X} \subset \mathbb{R}^n$, choose an initial level set ε_0 as,

$$\varepsilon_0 = \inf_{\rho}\{\varepsilon(P,\rho) : \mathcal{X} \subset \varepsilon(P,\rho)\} \tag{3.4}$$

We denote the value of ρ associated with ε_0 as ρ_0, and the corresponding values of ϵ, R and P as ϵ_0, R_0 and P_0 respectively. A simple approach to determining ε_0 and ρ_0 can also be found in [8]. More specifically, it is shown that the size of ε_0 grows monotonically as the parameter ρ grows. Hence, ε_0 and ρ_0 can be determined by a simple iteration procedure. Here we would like to note that, as explained in [8], increasing ρ indefinitely for exponentially unstable A will not result in an ε_0 that grows without bound.

To determine the inner level sets ε_i's, choose successivefully smaller ρ_i where $\rho_{i+1} < \rho_i$ for each $i = 1, 2, \cdots, N$. A simple choice of such ρ_i's is the geometric sequence of the form

$$\rho_i = \rho_0(\Delta\rho)^i, \quad i = 0, 1, 2, \cdots, N$$

where the ρ-reduction' factor $\Delta\rho \in (0, 1)$. (Consequently, the values of ϵ, R and P associated with each of these ρ_i's are denoted as ϵ_i, R_i and P_i respectively.) For a discussion of the choice of N and $\Delta\rho$, see [8].

As shown in [8], a critical property of such a sequence of level sets ε_i is that they are nested in the sense that $\varepsilon_i \subset \varepsilon_{i+1}$ for each $i = 0$ to $N - 1$.

3.2 Low-and-High Gain Feedback Design

Consider the linear dynamical system subject to actuator position saturation, input additive disturbances and uncertainties,

$$\dot{x} = Ax + B\text{sat}_\Delta(u + f(x) + d), \quad x(0) \in \mathcal{X} \subset \mathbb{R}^n, u \in \mathbb{R}^m \tag{3.5}$$

where the saturation function $\text{sat}_\Delta : \mathbb{R}^m \to \mathbb{R}^m$ is as defined in Section 2., the locally Lipschitz function $f : \mathbb{R}^n \to \mathbb{R}^m$ represents the input additive plant uncertainties and d the input-additive disturbance. The LHG feedback design for this system is given as follows. First, the level set ε_0 is determined as in the PLC design. Correspondingly, a state feedback law with a possibly low feedback gain is determined as,

$$u_L = -R_0^{-1}B'P_0x$$

A high gain state feedback is then constructed as,

$$u_H = -kR_0^{-1}B'P_0x, \quad k \geq 0$$

The final low-and-high gain state feedback is then given by a simple addition of the low and high gain feedbacks u_L and u_H, viz.,

$$u = -(1 + k)R_0^{-1}B'P_0x, \ k \geq 0$$

Here the design parameter k is referred to as the high gain parameter. As demonstrated in [7] the freedom in choosing the value of this high gain parameter can be utilized to achieve robust stabilization in the presence of input additive plant uncertainties $f(x)$ and input-additive disturbance rejection. Moreover, the transient speed for the states not in the range space of $B'P_0$ will increase as the value of k increases. To see this, let us consider the following Lyapunov function,

$$V_0(x) = x'P_0x$$

The evaluation of $\dot{V}$ along the trajectories of the closed-loop system in the absence of uncertainties and disturbances gives,

$$\dot{V} = -x'x - x'P_0BR_0^{-1}B'P_0x + 2x'P_0B[\text{sat}_\Delta(-(k+1)R_0^{-1}B'P_0x)$$
$$+ R_0^{-1}B'P_0x]$$

$$= -x'x - x'P_0BR_0^{-1}B'P_0x - 2\sum_{i=1}^{m} v_i[\text{sat}_{\Delta_i}((k+1)v_i) - v_i]$$

where we have denoted the ith element of $v = -R_0^{-1}B'P_0x$ as v_i. By the choice of P_0, it is clear that $|v_i| \leq \Delta_i$ and hence $-v_i[\text{sat}_{\Delta_i}((k+1)v_i) - v_i] \leq 0$, for each $i = 1$ to m. For any x not in the range space of $B'P_0$, that is $B'P_0x \neq 0$, then, for any i such that $v_i \neq 0$, $-v_i[\text{sat}_{\Delta_i}((k_2+1)v_i) - v_i] < -v_i[\text{sat}_{\Delta_i}((k_1+1)v_i) - v_i]$ if $k_2 > k_1$. However, for any x in the null space of $B'P_0$, $-v_i[\text{sat}_{\Delta_i}((k+1)v_i) - v_i] = 0$ for all i.

3.3 Combined PLC/LHG Design

In this subsection, we present the proposed combined PLC/LHG state feedback design for linear systems subject to actuator rate saturation (2.1).

The feedback control law design is carried out in the following three steps.

Step 1. Choose a pre-feedback

$$u = v + \bar{u} \tag{3.6}$$

Let $\tilde{x} = [x, v]'$. Then the system (2.1) under the above pre-feedback is given by,

$$\dot{\tilde{x}} = \tilde{A}\tilde{x} + \tilde{B}\text{sat}_\Delta(T_2u + f(\tilde{x}) + d), \ \tilde{x}(0) \in \mathcal{X} \times \mathcal{V} \subset \mathbb{R}^{n+m} \tag{3.7}$$

where

$$f(\tilde{x}) = [0 \ \ -T_1 + T_2]\tilde{x}$$

and,

$$\tilde{A} = \begin{bmatrix} A & B \\ 0 & 0 \end{bmatrix}, \quad \tilde{B} = \begin{bmatrix} 0 \\ I \end{bmatrix}$$

Assumption 2.1, i.e., the pair (A, B) is stabilizable, implies that $(\tilde{A}, \tilde{B})$ is stabilizable.

Step 2. Apply the PLC design to the system (3.7), and obtain a sequence of nested level sets $\varepsilon_0, \varepsilon_1, \cdots, \varepsilon_N$ (and correspondingly, the parameters $\epsilon_0, \epsilon_2, \cdots, \epsilon_N$) and a piecewise linear feedback law,

$$\bar{u} = \begin{cases} \bar{u}_i = -(\mu_2 T^*)^{-1} \tilde{R}_i^{-1} \tilde{B}' \tilde{P}_i \tilde{x} & \text{for } \tilde{x} \in \varepsilon_i \setminus \varepsilon_{i+1}, \ i = 0, \cdots, N-1 \\ \bar{u}_N = -(\mu_2 T^*)^{-1} \tilde{R}_N^{-1} \tilde{B}' \tilde{P}_N \tilde{x} & \text{for } \tilde{x} \in \varepsilon_N \end{cases}$$

(3.8)

Step 3. Design the LHG state feedback based on the PLC feedback law (3.8) and obtain the following combined final PLC/LHG feedback law,

$$u = \begin{cases} u_i = v - (k+1)(\mu_2 T^*)^{-1} \tilde{R}_i^{-1} \tilde{B}' \tilde{P}_i \tilde{x} \text{ for } \tilde{x} \in \varepsilon_i \setminus \varepsilon_{i+1}, \ i = 0, \cdots, N-1 \\ u_N = v - (k+1)(\mu_2 T^*)^{-1} \tilde{R}_N^{-1} \tilde{B}' \tilde{P}_N \tilde{x} \qquad \text{for } \tilde{x} \in \varepsilon_N \end{cases}$$

(3.9)

where $k \geq 0$ is a design parameter to be specified later.

3.4 Proof

In what follows, we will show as a theorem that the combined PLC/LHG state feedback law (3.9) solves Problem 2.1. The effectiveness of this feedback law in comparison with both the PLC and the LHG feedback laws will be demonstrated in the next Section.

Theorem 3.1. *Let Assumption 2.1 hold. Given the admissible data* (D, W_0), *there exists a* $k^*(D, W_0) \geq 0$ *such that, for all* $k \geq k^*$, *the combined PLC/LHG state feedback law (3.9) solves Problem 2.1. Moreover, if* $D = 0$, k^* *is independent of* W_0.

Proof. The proof is carried out in two steps. In the first step, we show that, for each $i = 0$ to $N-1$, there exits a $k_i^*(D) > 0$, such that for all $k \geq k_i^*$, in the presence of any d satisfying Assumption 2.3, all trajectories starting from $\varepsilon_i \setminus \varepsilon_{i+1}$ will remain in ε_i and enter into the inner level set ε_{i+1} in a finite time. This in turn implies that, for any $k \geq \max\{k_0, k_2, \cdots, k_{N-1}\}$, all the trajectories of the closed-loop system starting from $W \subset \varepsilon_0$ will enter the inner-most level set ε_N in a finite time. The second step of the proof is to show that: if $D = 0$, there exists a $k_N^* > 0$ such that, for all $k \geq k_N^*$, the equilibrium $\tilde{x} = 0$ of the closed-lop system is locally asymptotically stable with ε_N contained in its basin of attraction, and if $D \neq 0$, there exists a $k_N^*(D, W_0) > 0$ such that, for all $k \geq k_N^*$, all the trajectories of the closed-loop system starting from ε_N will remain in it and enter and remain in the set W_0 in a finite time. Throughout the proof, we will also notice that in the case that $D = 0$, all the k_i^*s are independent of D.

Once these two steps are completed, the proof of the theorem is then complete by taking $k^*(D, \mathcal{W}_0) = \max\{k_1^*, k_2^*, \cdots, k_N^*\}$.

We start by considering the closed-loop system for $\tilde{x} \in \varepsilon_i \setminus \varepsilon_{i+1}$, $i = 0$ to N,

$$
\begin{aligned}
\dot{\tilde{x}} &= \tilde{A}\tilde{x} + \tilde{B}\mathrm{sat}(-(k+1)T_2(\mu_2 T^*)^{-1}\tilde{R}_i^{-1}\tilde{B}'\tilde{P}_i\tilde{x} + f(\tilde{x}) + d) \\
&= (\tilde{A} - \tilde{B}\tilde{R}_i^{-1}\tilde{B}'\tilde{P}_i)\tilde{x} + \tilde{B}[\mathrm{sat}_\Delta(-(k+1)T_2(\mu_2 T^*)^{-1}\tilde{R}_i^{-1}\tilde{B}'\tilde{P}_i\tilde{x} + f(\tilde{x}) + d) \\
&\quad + \tilde{R}_i^{-1}\tilde{B}'\tilde{P}_i\tilde{x}]
\end{aligned}
$$

where $\varepsilon_{N+1} = \emptyset$.

We next pick the Lyapunov function,

$$
V_i = \tilde{x}'\tilde{P}_i\tilde{x} \tag{3.10}
$$

The evaluation of $\dot{V}_i$ along the trajectories of the closed-loop system in the presence of uncertainties and disturbances gives,

$$
\begin{aligned}
\dot{V}_i &= -\tilde{x}'\tilde{x} - \tilde{x}'\tilde{P}_i\tilde{B}\tilde{R}_i^{-1}\tilde{B}'\tilde{P}_i x \\
&\quad + 2\tilde{x}'\tilde{P}_i\tilde{B}[\mathrm{sat}_\Delta(-(k+1)T_2(\mu_2 T^*)^{-1}\tilde{R}_i^{-1}\tilde{B}'\tilde{P}_i\tilde{x} + f(\tilde{x}) + d) + \tilde{R}_i^{-1}\tilde{B}'\tilde{P}_i\tilde{x}] \\
&= -\tilde{x}'\tilde{x} - \tilde{x}'\tilde{P}_i\tilde{B}\tilde{R}_i^{-1}\tilde{B}'\tilde{P}_i\tilde{x} - 2\sum_{i=1}^{m} v_i[\mathrm{sat}_{\Delta_i}((k+1)\delta_i v_i + f_i + d_i) - v_i]
\end{aligned}
$$

where we have denoted the ith elements of $v = -\tilde{R}_i^{-1}\tilde{B}'\tilde{P}_i x$, $f(\tilde{x})$ and d respectively as v_i, f_i and d_i, and $\delta_i = \tau_{2i}/\mu_{2i}\tau_i^* \geq 1$.

By the construction of ε_i, it is clear that $|v_i| \leq \Delta_i$ for all $\tilde{x} \in \varepsilon_i$. Also note that

$$
\|f(\tilde{x})\| \leq (\|\mu_2 - \nu_1\| + \|\mu_1 - \nu_2\|)\|T^*\|\|\tilde{x}\| = \beta\|x\| \tag{3.11}
$$

Hence we have,

$$
|[(k+1)\delta_i - 1]v_i| \geq |f_i + d_i| \Longrightarrow -2v_i[\mathrm{sat}_{\Delta_i}((k+1)\delta_i v_i + f_i + d_i) - v_i] \leq 0
$$

and,

$$
\begin{aligned}
|[(k+1)\delta_i - 1]v_i| < |f_i + d_i| \Longrightarrow &-2v_i[\mathrm{sat}_{\Delta_i}((k+1)\delta_i v_i + f_i + d_i) - v_i] \\
&= -2v_i[\mathrm{sat}_{\Delta_i}((k+1)\delta_i v_i + f_i + d_i) \\
&\quad -\mathrm{sat}_{\Delta_i}(v_i)] \\
&\leq \frac{2|f_i + d_i|}{(k+1)\delta_i - 1}|[(k+1)\delta_i - 1]v_i + f_i + d_i| \\
&\leq \frac{8(f_i^2 + d_i^2)}{k}
\end{aligned}
$$

where we have used the fact that,

$$
|\mathrm{sat}_{\Delta_i}(v_1) - \mathrm{sat}_{\Delta_i}(v_2)| \leq |v_1 - v_2|, \quad \forall v_1, v_2 \in \mathbb{R}
$$

Hence, we can conclude that, for all $\tilde{x} \in \varepsilon_i \setminus \varepsilon_{i+1}$, $i = 0$ to N,

$$\dot{V}_i \leq -\tilde{x}'\tilde{x} + \frac{8\beta^2}{k}\tilde{x}'\tilde{x} + \frac{8\|d\|^2}{k} \leq -\left(1 - \frac{8}{k}\right)\tilde{x}'\tilde{x} + \frac{8D^2}{k} \qquad (3.12)$$

To complete the first step of the proof, for each $i = 0$ to $N - 1$, we let

$$k_i^*(D) = \max\left\{16\beta^2, \frac{17\lambda_{\max}(P_{i+1})D^2}{\rho_{i+1}}\right\}$$

where β is as defined in (3.11). We then have that, for all $k \geq k_i^*$,

$$\dot{V}_i \leq -\frac{1}{2\lambda_{\max}(P_{i+1})}\left(V_{i+1} - \frac{16\lambda_{\max}(P_{i+1})D^2}{k}\right) \qquad (3.13)$$

and hence, for each $i = 0$ to $N - 1$,

$$\dot{V}_i < 0, \quad \forall \tilde{x} \in \varepsilon_i \setminus \varepsilon_{i+1} \qquad (3.14)$$

which, together with the fact that ε_{i+1} is strictly inside ε_i, shows that all the trajectories of the closed-loop system starting from $\varepsilon_i \setminus \varepsilon_{i+1}$ will remain in ε_i and enter the level set ε_{i+1} in a finite time. We note that in the case of $D = 0$, each of the $k_i^*(D)$, $i = 0$ to $N - 1$, as chosen above is independent of D.

We next proceed with the second step of the proof. In the case that $D = 0$, choose $k_N^* = 16\beta^2$, independent of D. It then follows from (3.12) that

$$\dot{V}_N \leq -\frac{1}{2}\tilde{x}'\tilde{x} \qquad (3.15)$$

and hence the equilibrium $\tilde{x} = 0$ of the closed-loop system is locally asymptotically stable with ε_N contained in its basin of attraction. In the case that $D \neq 0$, let $\rho_{N+1} \in (0, \rho_N]$ be such that $\varepsilon(P_N, \rho_{N+1}) \subset \mathcal{W}_0$. The existence of such a ρ_{N+1} is due to the fact that $\mathcal{W}_0$ contains the origin of the state space as an interior point. Choose $k_N^*(D, \mathcal{W}_0)$ as follows,

$$k_N^*(D, \mathcal{W}_0) = \max\left\{16\beta^2, \frac{17\lambda_{\max}(P_N)D^2}{\rho_{N+1}}\right\}$$

We then have that, for all $k \geq k_N^*$,

$$\dot{V}_N \leq -\frac{1}{2\lambda_{\max}(P_N)}\left(V_N - \frac{16\lambda_{\max}(P_N)D^2}{k}\right) \qquad (3.16)$$

and hence,

$$\dot{V}_N < 0, \quad \forall \tilde{x} \in \varepsilon_N \setminus \varepsilon(P_N, \rho_{N+1}) \qquad (3.17)$$

which shows that all trajectories of the closed-loop system starting from ε_N will remain in ε_N and enter the set $\mathcal{W}_0 \supset \varepsilon(P_N, \rho_{N+1})$ in a finite time and remain in it thereafter. $\square$

4. Stabilizing Feedback Design for an F-16 Aircraft

In this section, the applicability and effectiveness of the proposed combined PLC/LHG design algorithm is demonstrated with an F-16 fighter aircraft derivative. At the flight condition corresponding to an altitude of 10,000 feet and a Mach number of 0.7, the second order pitch plane dynamics (short period) of this aircraft is given by ([5]),

$$
\begin{cases}
\begin{bmatrix} \dot{\alpha} \\ \dot{q} \end{bmatrix} = \begin{bmatrix} -1.1500 & 0.9937 \\ 3.7240 & -1.2600 \end{bmatrix} \begin{bmatrix} \alpha \\ q \end{bmatrix} + \begin{bmatrix} -0.1770 \\ -19.5000 \end{bmatrix} \delta, & |\alpha(0)| \le 0.1 \text{rad}, \\[4pt]
& |q(0)| \le 1 \text{rad/sec} \\[4pt]
\dot{\delta} = \text{sat}_1(-T_1\delta + T_2 u + d), & |\delta(0)| \le 0.1 \text{rad}
\end{cases}
$$

where α, q and δ are respectively the trim value of the angle of attack, pitch rate and stabilizer deflection. Here the actuator is rate limited to ± 1.0 rad/sec. The constants T_1 and T_2 have a same nominal value of 20 1/sec, which corresponds to a nominal actuator bandwidth of 20 rad/sec. However, the actual values of T_1 and T_2 vary and lie within $\pm 5\%$ of the nominal value, i.e., $\mu_1 = \mu_2 = 0.95$ and $\nu_1 = \nu_2 = 1.05$.

It is straightforward to see that the above system is in the form of (2.1). Our goal here is to design an effective robust stabilizing feedback law that will also be able to reject the actuator disturbance d. The feedback law we are to design is the combined PLC/LHG feedback law proposed in the previous section.

Following the design procedure given in Section 3.3, we first find,

$$
\rho_0 = 1.2335, \quad \epsilon_0 = 5.0805, \quad P_0 = \begin{bmatrix} 1.0456 & 0.4868 & -3.3592 \\ 0.4868 & 0.4185 & -2.1537 \\ -3.3592 & -2.1537 & 20.9250 \end{bmatrix}
$$

Choosing $N = 5$ and $\Delta\rho = 0.6$, we obtain a feedback law of the form (3.9) with,

$$
\rho_1 = 0.7401, \quad \epsilon_1 = 3.3505, \quad P_1 = \begin{bmatrix} 0.9130 & 0.4107 & -2.5622 \\ 0.4107 & 0.3737 & -1.7118 \\ -2.5622 & -1.7118 & 15.1680 \end{bmatrix}
$$

$$
\rho_2 = 0.4441, \quad \epsilon_2 = 2.2150, \quad P_2 = \begin{bmatrix} 0.8083 & 0.3486 & -1.9619 \\ 0.3486 & 0.3357 & -1.3696 \\ -1.9619 & -1.3696 & 11.0484 \end{bmatrix}
$$

$$
\rho_3 = 0.2664, \quad \epsilon_3 = 1.4675, \quad P_3 = \begin{bmatrix} 0.7252 & 0.2975 & -1.5072 \\ 0.2975 & 0.3030 & -1.1021 \\ -1.5072 & -1.1021 & 8.0825 \end{bmatrix}
$$

$$
\rho_4 = 0.1599, \quad \epsilon_4 = 0.9741, \quad P_4 = \begin{bmatrix} 0.6592 & 0.2552 & -1.1613 \\ 0.2552 & 0.2746 & -0.8912 \\ -1.1613 & -0.8912 & 5.9355 \end{bmatrix}
$$

$$\rho_5 = 0.0959, \quad \epsilon_5 = 0.6477, \quad P_5 = \begin{bmatrix} 0.6066 & 0.2199 & -0.8972 \\ 0.2199 & 0.2496 & -0.7235 \\ -0.8972 & -0.7235 & 4.3736 \end{bmatrix}$$

We note here again that this control law reduces to an LHG feedback control law if $N = 0$ and to a PLC law if $k = 0$ and $\mu_1 = \mu_2 = \nu_1 = \nu_2 = 1$.

Figs. 4.1 and 4.2 are some simulation results. In these simulations, $T_1 = 21$, $T_2 = 19$, $d = 5\sin(10t + 3)$. These simulations show good utilization of the available actuator rate capacity, and the degree of disturbance rejection increases as the value of the parameter k increases. Higher inaccuracy in the actuator parameter can be tolerated by further increasing the value of the design parameter k.

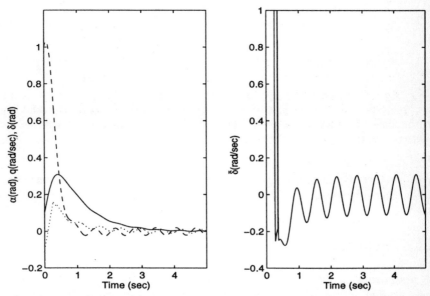

Fig. 4.1. Combined PLC/LHG design ($N = 5$, $k = 64$). $T_1 = 21$, $T_2 = 19$, $d = 5\sin(10t + 3)$. The left plot: states of the plant and the actuator, the solid line is the angle of attack α, the dashed line the pitch rate q, and the dotted line the deflection δ; the right plot: the actuator rate $\dot{\delta}$.

We next compare the performance of the combined PLC/LHG design with both the PLC and the LHG design. To compare with the PLC laws, we set $\mu_2 = 1$ and $k = 0$ in (3.9). Simulation (Fig. 4.3) shows that even in the absence of a disturbance, the aircraft goes unstable due to the presence of the 15% inaccuracy in the actuator parameters T_1 and T_2. To compare with the LHG feedback law, we set $N = 0$ in (3.9), resulting a feedback law with constant gain $-(\mu_2 T^*)^{-1}\tilde{R}^{-1}\tilde{B}'\tilde{P}_0$. Simulations (Fig. 4.4) shows that the combined PLC/LHG feedback law results in much better transient

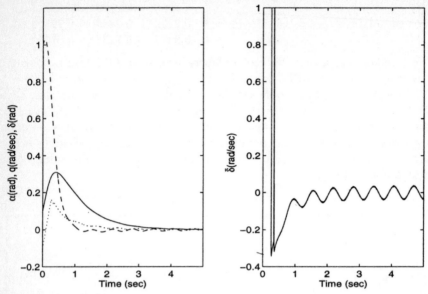

Fig. 4.2. Combined PLC/LHG design ($N = 5$, $k = 200$). $T_1 = 21$, $T_2 = 19$, $d = 5\sin(10t + 3)$. The left plot: states of the plant and the actuator, the solid line is the angle of attack α, the dashed line the pitch rate q, and the dotted line the deflection δ; the right plot: the actuator rate $\dot{\delta}$.

performance. By increasing the value of N, the transience performance can be further improved.

5. Conclusions

A stabilizing feedback control law design method for linear systems with rate-limited actuators is proposed. This design method is based on two design techniques recently developed for linear systems with position limited actuators and takes advantages of the both design techniques, while avoiding their disadvantages. Application of the proposed design to an F-16 fighter aircraft demonstrates the applicability, robustness and effectiveness of the novel control law.

Following the development of [7], high gain observer based output feedback results can also be developed in a straightforward manner.

References

1. J.M. Berg, K.D. Hammett, C.A. Schwartz, and S.S. Banda, "An analysis of

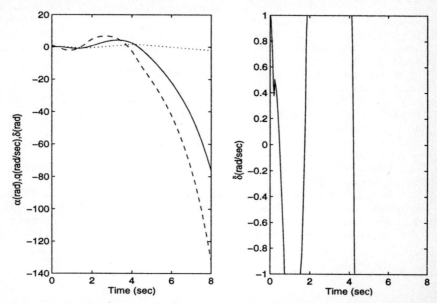

Fig. 4.3. PLC Design $(k = 0)$. $T_1 = 17$, $T_2 = 23$, $d = 0$. The left plot: states of the plant and the actuator, the solid line is the angle of attack α, the dashed line the pitch rate q, and the dotted line the deflection δ; the right plot: the actuator rate $\dot{\delta}$.

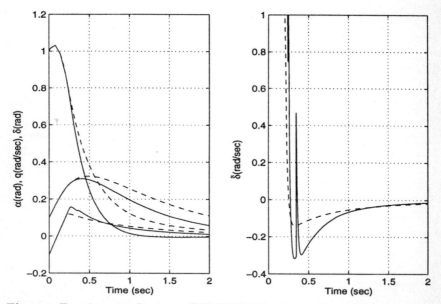

Fig. 4.4. Transience performance: PLC/LHG $(N = 5, k = 10$, solid lines) vs. LHG $(N = 0, k = 10$, dashed lines). $T_1 = 21$, $T_2 = 19$, $d = 0$.

the destabilizing effect of daisy chained rate-limited actuators," *IEEE Trans. Control Sys. Tech.*, Vol. 4, No. 2, 1996.

2. D.S. Bernstein and A.N. Michel, "A chronological bibliography on saturating actuators," *International Journal of Robust and Nonlinear Control*, Vol.5, No.5, pp.375-380, 1995.

3. M.A. Dornheim, "Report pinpoints factors leading to YF-22 crash," *Aviation Week Space Technol.*, pp. 53-54, Nov. 9, 1992.

4. J.M. Lenorovitz, "Gripen control problems resolved through in-flight, ground simulations," *Aviation Week Space Technol.*, pp. 74-75, June 18, 1990.

5. M. Pachter, P.R. Chandler and M. Mears, "Reconfigurable tracking control with saturation," *AIAA Journal of Guidance, Control and Dynamics*, Vol. 18, pp 1016-1022, 1995.

6. C.A. Shifrin, "Gripen likely to fly again soon," *Aviation Week Space Technol.*, pp. 72, Aug. 23, 1993.

7. A. Saberi, Z. Lin and A.R. Teel, "Control of linear systems subject to input saturation," *IEEE Trans. Auto. Contr.*, Vol. 41, No.3, pp. 368-378, 1996.

8. G.F. Wredenhagen and P.R. Belanger, "Piecewise-linear LQ control for systems with input constraints," *Automatica*, vol. 30, pp. 403-416, 1994.

Lecture Notes in Control and Information Sciences

Edited by M. Thoma

Vol. 204: Takahashi, S.; Takahara, Y.
Logical Approach to Systems Theory
192 pp. 1995 [3-540-19956-X]

Vol. 205: Kotta, U.
Inversion Method in the Discrete-time
Nonlinear Control Systems Synthesis
Problems
168 pp. 1995 [3-540-19966-7]

Vol. 206: Aganovic, Z.;.Gajic, Z.
Linear Optimal Control of Bilinear Systems
with Applications to Singular Perturbations
and Weak Coupling
133 pp. 1995 [3-540-19976-4]

Vol. 207: Gabasov, R.; Kirillova, F.M.;
Prischepova, S.V.
Optimal Feedback Control
224 pp. 1995 [3-540-19991-8]

Vol. 208: Khalil, H.K.; Chow, J.H.;
Ioannou, P.A. (Eds)
Proceedings of Workshop on Advances
inControl and its Applications
300 pp. 1995 [3-540-19993-4]

Vol. 209: Foias, C.; Özbay, H.;
Tannenbaum, A.
Robust Control of Infinite Dimensional
Systems: Frequency Domain Methods
230 pp. 1995 [3-540-19994-2]

Vol. 210: De Wilde, P.
Neural Network Models: An Analysis
164 pp. 1996 [3-540-19995-0]

Vol. 211: Gawronski, W.
Balanced Control of Flexible Structures
280 pp. 1996 [3-540-76017-2]

Vol. 212: Sanchez, A.
Formal Specification and Synthesis of
Procedural Controllers for Process Systems
248 pp. 1996 [3-540-76021-0]

Vol. 213: Patra, A.; Rao, G.P.
General Hybrid Orthogonal Functions and
their Applications in Systems and Control
144 pp. 1996 [3-540-76039-3]

Vol. 214: Yin, G.; Zhang, Q. (Eds)
Recent Advances in Control and Optimization
of Manufacturing Systems
240 pp. 1996 [3-540-76055-5]

Vol. 215: Bonivento, C.; Marro, G.;
Zanasi, R. (Eds)
Colloquium on Automatic Control
240 pp. 1996 [3-540-76060-1]

Vol. 216: Kulhavý, R.
Recursive Nonlinear Estimation: A Geometric
Approach
244 pp. 1996 [3-540-76063-6]

Vol. 217: Garofalo, F.; Glielmo, L. (Eds)
Robust Control via Variable Structure and
Lyapunov Techniques
336 pp. 1996 [3-540-76067-9]

Vol. 218: van der Schaft, A.
L_2 Gain and Passivity Techniques in Nonlinear
Control
176 pp. 1996 [3-540-76074-1]

Vol. 219: Berger, M.-O.; Deriche, R.;
Herlin, I.; Jaffré, J.; Morel, J.-M. (Eds)
ICAOS '96: 12th International Conference on
Analysis and Optimization of Systems -
Images, Wavelets and PDEs:
Paris, June 26-28 1996
378 pp. 1996 [3-540-76076-8]

Vol. 220: Brogliato, B.
Nonsmooth Impact Mechanics: Models,
Dynamics and Control
420 pp. 1996 [3-540-76079-2]

Vol. 221: Kelkar, A.; Joshi, S.
Control of Nonlinear Multibody Flexible Space
Structures
160 pp. 1996 [3-540-76093-8]

Vol. 222: Morse, A.S.
Control Using Logic-Based Switching
288 pp. 1997 [3-540-76097-0]

Vol. 223: Khatib, O.; Salisbury, J.K.
Experimental Robotics IV: The 4th International
Symposium, Stanford, California,
June 30 - July 2, 1995
596 pp. 1997 [3-540-76133-0]